"The author of *Aphids Unveiled* uses a funny and engaging style to share info about aphids, from their beginnings to why they bug farmers and gardeners. With the author being an insect expert and professor, the book is both informative and fun. The mix of science knowledge and storytelling make it interesting for a wide audience, from people into gardening to those who just like learning about nature. *Aphids Unveiled* fills a gap by giving us a good and enjoyable look into these often ignored but important insects."

Professor Ahmet Bayram, *Applied Agricultural Entomologist,*
Dicle University, Turkey

W0018307

Aphids Unveiled

Aphids Unveiled: The Saga of Nature's Most Irritating Insects explores the curious journey of these tiny insects from humble beginnings to becoming the most obnoxious pests for farmers and home gardeners alike. With a witty narrative style and a touch of humor, the book aims to engage readers with the quirky behaviors, social structures, and survival strategies of aphids, all while unraveling the reasons behind their impact on agriculture and horticulture.

Amanda Rose Newton brings over a decade of experience in both the horticulture and pest control worlds to write for a broad audience, from gardening enthusiasts seeking effective pest management strategies to science enthusiasts interested in a lighthearted exploration of the natural world. As an entomologist and professor, her writing uniquely blends expertise with a passion for storytelling, ensuring *Aphids Unveiled* is not only informative but also an entertaining read.

Aphids Unveiled

The Saga of Nature's Most Irritating Insects

Amanda Rose Newton

CRC Press
Taylor & Francis Group
Boca Raton London New York

CRC Press is an imprint of the
Taylor & Francis Group, an **informa** business

Designed cover image: Shutterstock ID: 1943731147

First edition published 2025
by CRC Press
2385 NW Executive Center Drive, Suite 320, Boca Raton FL 33431

and by CRC Press
4 Park Square, Milton Park, Abingdon, Oxon, OX14 4RN

CRC Press is an imprint of Taylor & Francis Group, LLC

ISBN: 978-1-032-88441-7 (hbk)
ISBN: 978-1-032-88346-5 (pbk)
ISBN: 978-1-003-53778-6 (ebk)

DOI: 10.1201/9781003537786

Typeset in Sabon
by KnowledgeWorks Global Ltd.

To Dad

Contents

About the author

Amanda Rose Newton is a passionate entomologist and writer known for her unique ability to blend science with humor. With a love for all things creepy-crawly, Amanda has spent years studying the intricate world of insects, particularly those tiny yet notorious garden pests—aphids.

In addition to her writing, Amanda holds a degree in entomology and has spent years working in various capacities as a researcher, educator, and consultant. She is known for her approachable teaching style, making complex scientific concepts accessible and enjoyable. Whether through her writing, public speaking, or educational outreach, Amanda's mission is to inspire a greater appreciation for the tiny yet vital insects that play such a crucial role in our ecosystems.

When she's not writing or studying insects, Amanda enjoys gardening, where she continues to observe and appreciate the very creatures she writes about, always finding new material for her next humorous take on the natural world.

Preface

Welcome to the tiny, perplexing, and surprisingly comical world of aphids. Whether you're an experienced gardener, an insect enthusiast, or simply someone who's ever noticed a mysterious cluster of tiny green dots on your favorite plants, this book is for you. "Aphids Unveiled" is not your typical garden guide—it's a lighthearted journey into the lives of these minute marvels that will leave you both educated and entertained.

Aphids are often viewed as the ultimate garden villains—those minuscule, sap-sucking insects that can turn a lush, vibrant plant into a wilting mess seemingly overnight. But behind their bad reputation lies a world of complex relationships, bizarre behaviors, and even a few redeeming qualities (yes, you read that right). Through the pages of this book, we'll peel back the layers of misunderstanding and take a closer look at what makes aphids such fascinating creatures.

In writing *Aphids Unveiled: The Saga of Nature's Most Irritating Insects*, I aimed to create a book that is equal parts informative and amusing. As an entomologist with a love for storytelling, I've found that humor is often the best way to make the smallest, most overlooked details of nature spring to life. After all, who said learning about insects can't be fun?

Throughout this book, you'll discover the curious biology of aphids, their peculiar life cycles, and their impact on our gardens. You'll meet the characters who call aphids their prey, and you'll learn why aphids aren't always the bad guys they're made out to be. Along the way, you might just find yourself laughing out loud—after all, the aphid's world is full of unexpected twists, turns, and a few too many puns.

So grab your magnifying glass, settle in, and prepare to view aphids in a whole new light. Let's embark on this tiny adventure together, one that promises to be as enlightening as it is entertaining.

Glossary

A

Aphid: Small sap-sucking insects, members of the superfamily Aphidoidea, known for their rapid reproduction and ability to cause significant damage to crops and plants.

Aphid-Alien Connection: A humorous theory that aphids' unique reproductive strategies are evidence of extraterrestrial life.

Aphid-Resistant Plants: Crop varieties that possess natural defenses against aphid infestation, either through genetic modification or selective breeding.

Artificial Intelligence (AI): Advanced computer systems capable of performing tasks that typically require human intelligence, such as visual perception, speech recognition, decision-making, and language translation. In agriculture, AI can predict pest outbreaks and suggest timely interventions.

Alarm Pheromones: Chemical signals released by aphids when attacked, causing others to flee.

B

Bacteriocytes: Specialized cells in aphids that house symbiotic bacteria, providing essential nutrients.

Beneficial Insects: Insects that provide natural pest control by preying on or parasitizing harmful insect species.

Biological Control: The use of natural predators, parasites, or pathogens to manage pest populations.

Biopesticides: Pesticides derived from natural materials such as animals, plants, bacteria, and certain minerals. Used to control pests in an environmentally friendly way.

Buchnera aphidicola: Symbiotic bacteria that reside within aphids and provide essential nutrients, allowing them to thrive on a diet of plant sap.

C

Chemical Insecticides: Synthetic chemicals used to kill or repel insects.

Climate Change: Long-term changes in temperature and weather patterns, often attributed to human activity. Can impact aphid populations and their behavior.

Colony Dynamics: The interactions and organizational structure within a colony of aphids, including social hierarchy and communication.

Companion Planting: Planting different crops in proximity for pest control, pollination, and maximizing use of space.

D

Dropping: A defense behavior in which aphids fall off plants to escape predators.

E

Endosymbionts: Microorganisms that live inside another organism (host), forming a symbiotic relationship.

Entomopathogenic Fungi: Fungi that infect and kill insects, used as biological control agents.

Environmental Adaptation: The ability of organisms to adjust to environmental changes to survive and thrive.

F

Folklore: Traditional beliefs, customs, and stories passed through generations, often featuring aphids in various symbolic roles.

G

Genetic Engineering: The manipulation of an organism's DNA to produce desired traits, such as pest resistance in crops.

Green Peach Aphid: A common species of aphid known for its ability to transmit plant viruses and rapidly reproduce.

H

Honeydew: A sugary substance excreted by aphids that attracts ants and can promote the growth of sooty mold.

I

Integrated Pest Management (IPM): A sustainable approach to managing pests that combines biological, chemical, and cultural methods.

Instars: The stages between molts in the lifecycle of an insect.

M

Molting: The process of shedding the exoskeleton to allow for growth in insects.

Mutualism: A symbiotic relationship where both parties benefit.

N

Neem Oil: A natural pesticide derived from the seeds of the neem tree, used to control a variety of pests.

P

Parthenogenesis: A form of asexual reproduction where an organism produces offspring without fertilization.

Photoperiod: The length of day and night, which can influence biological processes in organisms.

Precision Agriculture: Farming practices that use technology to monitor and manage crops with high precision.

R

Reflective Mulch: A type of mulch that reflects light, deterring pests like aphids from landing on plants.

S

Sticky Traps: Devices coated with a sticky substance used to trap flying insects.

Sustainable Agriculture: Farming practices that maintain and improve environmental health, economic profitability, and social equity.

T

Tactile Signals: Communication through touch, used by aphids to transmit information.

W

Winged Aphids: Aphids that develop wings to migrate to new plants when conditions become unfavorable.

Y

Yellow Sticky Traps: Traps used to catch flying aphids and other pests, helping to monitor and reduce their populations.

CHAPTER 1

The rise of the tiny tyrants

Introduction

Aphids, those tiny green (or sometimes black, yellow, or pink) sap-sucking insects, have a history as rich as their appetites. These minuscule marauders have been around for millions of years, long before humans even thought of planting crops or complaining about pests. Yet their story is inextricably linked to the rise of agriculture. From their modest beginnings as small blips on the evolutionary radar to their current status as one of the most persistent agricultural pests, aphids have evolved and adapted in remarkable ways. This chapter delves into the evolutionary journey of these tiny tyrants, shedding light on their origins, early encounters with humans, and the significant impact they have had on agriculture throughout history.

The origins of aphids

Imagine a world teeming with giant ferns and early conifer forests. This is where our story begins, some 280 million years ago, in the Permian period. Aphids belong to the superfamily Aphidoidea, part of the order Hemiptera, commonly known as true bugs. Fossil records indicate that aphids

Timeline of Aphid Evolution

- **280 million years ago:** Earliest aphid fossils found in the Permian period.
- **200 million years ago:** Diversification of aphid species during the Mesozoic era.
- **65 million years ago:** Survival and adaptation through the Cretaceous-Paleogene extinction event.
- **2 million years ago:** Emergence of aphids as significant pests in early agricultural societies.

DOI: 10.1201/9781003537786-1

were already equipped with their signature piercing-sucking mouthparts, an adaptation that has been key to their survival and success. These ancient aphids were the original "plant vampires," silently sipping on the primordial sap of prehistoric plants.

Aphids and early agriculture

Fast forward to the dawn of human agriculture. As humans transitioned from hunter-gatherer societies to settled agricultural communities, they began to cultivate plants on a larger scale. This shift in human activity provided aphids with abundant food sources and new opportunities for population growth. Early farmers likely encountered aphids as they noticed these insects infesting their crops, leading to the first recorded instances of pest management.

The evolutionary adaptations of aphids

Aphids have evolved several unique adaptations that have contributed to their success as pests. Their reproductive strategies, including parthenogenesis (asexual reproduction) and viviparity (giving birth to live young), allow them to rapidly increase their populations. Additionally, their ability to produce multiple generations in a single season and their development of specialized mouthparts for feeding on plant sap have made them particularly formidable.

Aphids also exhibit complex life cycles, often involving host alternation. This means that some species can switch between different host plants during their life cycle, further increasing their chances of survival and spread. This adaptability has enabled aphids to exploit a wide range of plants, including many agricultural crops.

> ## Aphids in Ancient Agriculture
>
> Historical records from ancient civilizations, such as the Egyptians and the Greeks, mention the presence of small, sap-sucking insects on crops. While these records do not specifically name aphids, the descriptions align with the characteristics of aphid infestations. These early mentions highlight the long-standing battle between humans and aphids.

> ## Aphid Reproductive Strategies
>
> - **Parthenogenesis:** Female aphids can produce offspring without mating, leading to rapid population growth.
> - **Viviparity:** Aphids give birth to live young, bypassing the egg stage and speeding up reproduction.
> - **Host Alternation:** Some aphid species can switch between different host plants, increasing their adaptability.

Case study: The phylloxera crisis in European vineyards

One of the most significant historical impacts of aphids on agriculture is the Phylloxera crisis in European vineyards during the 19th century. Phylloxera, an aphid-like insect, devastated vineyards across Europe, particularly in France, leading to widespread economic and social upheaval.

Phylloxera (*Daktulosphaira vitifoliae*) originated in North America and was inadvertently introduced to Europe through the importation of American grapevines. Unlike European grapevines, which had no resistance to Phylloxera, American grapevines had co-evolved with the pest and developed natural defenses.

The introduction of Phylloxera to Europe led to the destruction of vast expanses of vineyards. The pest attacked the roots of the grapevines, causing them to wither and die. The crisis prompted a desperate search for solutions, including grafting European grapevines onto resistant American rootstocks, which ultimately saved the European wine industry.

First encounters with agriculture

The first recorded encounters between humans and aphids in agricultural settings provide fascinating insights into the challenges faced by early farmers. Ancient texts and archaeological findings reveal that aphid infestations were a common problem in early agricultural societies.

In ancient Egypt, farmers dealt with aphids on their crops by using various pest control methods, including the application of natural repellents and the introduction of beneficial insects. Similarly, the Greeks and Romans documented aphid infestations and their attempts to manage these pests through both cultural practices and early chemical controls.

Aphids as agricultural pests

As agriculture expanded and became more intensive, the impact of aphids on crop production grew more pronounced. Aphids are particularly problematic because they not only feed on plant sap but also transmit plant viruses, which can cause significant damage to crops.

The economic impact of aphid infestations on agriculture is substantial. Aphids can reduce crop yields, affect the quality of produce, and increase the costs of pest management. Their ability to rapidly reproduce and form large colonies makes them challenging to control, and their presence can lead to secondary pest issues as well.

The modern challenge

Today, aphids remain a major concern for farmers and gardeners worldwide. Modern agricultural practices and global trade have contributed to the spread of aphids and their associated plant viruses. Integrated pest

Figure 1.1 Oleander aphid. (Shutterstock ID: 2151158093)

management (IPM) strategies, which combine biological, cultural, and chemical controls, are essential for managing aphid populations effectively.

Conclusion

The rise of aphids as agricultural pests is a testament to their remarkable adaptability and evolutionary success. From their ancient origins to their impact on modern agriculture, aphids have proven to be resilient and formidable adversaries. Understanding their evolutionary history and the factors that have contributed to their emergence as pests is crucial for developing effective management strategies and ensuring the sustainability of our agricultural systems.

As we continue to face the challenges posed by aphids, it is essential to draw on the lessons of history and leverage modern scientific advancements to develop innovative solutions. By doing so, we can mitigate the impact of these tiny tyrants and safeguard the future of agriculture.

Glossary of Terms

Agricultural Pests: Organisms that damage crops and reduce agricultural productivity.

Aphidoidea: A superfamily of insects in the order Hemiptera, commonly known as aphids.

Hemiptera: An order of insects also known as true bugs, characterized by their piercing-sucking mouthparts.

Host Alternation: A life cycle strategy where an organism alternates between different host species during different stages of its life cycle.

Integrated Pest Management (IPM): An approach to pest control that combines biological, cultural, and chemical methods to manage pest populations effectively.

Parthenogenesis: A form of asexual reproduction where an organism can produce offspring without mating.

Phylloxera (*Daktulosphaira vitifoliae*): An aphid-like pest that devastated European vineyards in the 19th century.

Piercing-sucking mouthparts: Specialized mouthparts used by certain insects to pierce plant tissues and suck out sap.

Plant Viruses: Pathogens that infect plants and can be transmitted by insect vectors such as aphids.

Sap-sucking: Feeding behavior of certain insects that involves extracting sap from plants.

Viviparity: The condition where organisms give birth to live young, bypassing the egg stage.

Bibliography

Blackman, R. L., & Eastop, V. F. (2000). *Aphids on the World's Crops: An Identification and Information Guide*. Wiley.

Dixon, A. F. G. (1998). *Aphid Ecology: An Optimization Approach*. Chapman & Hall.

Van Emden, H. F., & Harrington, R. (2007). *Aphids as Crop Pests*. CABI.

CHAPTER 2

Aphid anthropology

Introduction

If you've ever had the pleasure of meeting an aphid up close and personal (and by pleasure, I mean frustration), you might think of them as solitary little sap-suckers, each focused on its own sugary feast. But beneath this seemingly solitary existence lies a bustling, intricate society. Welcome to the social world of aphids, where cooperation, communication, and communal living are the order of the day. Think of it as a tiny version of our own societies, but with more legs and a lot more sap.

The social structure of aphid colonies

Aphids, it turns out, have a social structure that would make any ant colony jealous. These little insects live in colonies that can range from a few individuals to thousands, all working together in a well-coordinated community. Each colony is a mix of winged and wingless individuals, soldiers, and reproductive members, each with specific roles to play.

In the world of aphids, size doesn't matter—it's all about numbers and cooperation. Aphids can form dense colonies, sometimes covering entire plant stems and leaves. The members of these colonies are organized in a way that maximizes their survival and reproductive success.

Aphids exhibit a form of social organization known as "subsociality." While they do not form the highly complex societies seen in ants or bees, their colonies are structured and cooperative. Subsociality involves some level of parental care and cooperative behaviors among colony members, which is significant for such small insects.

DOI: 10.1201/9781003537786-2

In these colonies, some aphids take on defensive roles while others focus on reproduction. Winged aphids often act as scouts, searching for new host plants when resources become scarce. This division of labor ensures that the colony can adapt to changing environmental conditions and threats.

Case study: The soldier aphids

Some aphid species have specialized "soldier" aphids that defend the colony against predators. Take, for example, the gall-forming aphid *Pemphigus spyrothecae*. The soldiers in this species are typically first-instar nymphs, armed with fierce jaws ready to fend off ladybird larvae and other threats. These tiny warriors sacrifice their lives to protect their siblings, a selfless act that ensures the colony's survival.

These soldiers are nature's version of kamikaze pilots, ready to give their all for the greater good of the colony. It's a harsh world out there, and aphids have evolved to meet the challenge head-on—or in this case, mandibles-on.

The evolution of soldier aphids is a remarkable example of kin selection, where individuals perform altruistic acts that benefit their relatives. By protecting the colony, soldier aphids increase the survival chances of their genetically similar siblings, thus ensuring the propagation of their shared genes.

The presence of soldier aphids can significantly reduce predation rates, as these defenders actively seek out and attack intruding predators. Their effectiveness in defense highlights the adaptive value of having specialized castes within aphid colonies.

Communication: The aphid way

Communication is key in any society, and aphids are no different. They use a variety of methods to keep in touch, alert each other to danger, and coordinate their activities. One of the most fascinating methods is the use of alarm pheromones. When an aphid is attacked, it releases a chemical signal that causes nearby aphids to scatter—a kind of "every bug for itself" strategy.

Aphids also use tactile signals, communicating through touch. They tap each other with their antennae, a bit like Morse code but without the beeps. This allows them to convey important information quickly and efficiently.

Aphid Communication Methods

- **Alarm Pheromones:** Chemical signals released when an aphid is attacked, causing others to flee.
- **Honeydew:** A sugary substance excreted by aphids that attracts ants. The ants protect the aphids in exchange for this sweet treat.
- **Tactile Signals:** Aphids use their antennae to tap and touch each other, transmitting information through these gentle touches.

Figure 2.1 Honeydew. (Shutterstock ID: 1943731147)

Aphids excel in chemical and tactile communication, but they also use visual signals to interact. For example, winged aphids can alert their wingless counterparts when it's time to take flight, which is crucial during an attack when a quick escape is necessary.

In some species, aphids change color in response to environmental cues like the presence of predators or temperature fluctuations. This color change acts as a warning to other aphids, prompting them to take defensive measures.

When an aphid is attacked, it releases alarm pheromones, primarily composed of compounds like (E)-β-farnesene, from specialized glands. This rapid communication system ensures that other aphids can quickly respond to threats, minimizing casualties within the colony.

The production of honeydew serves multiple purposes. It attracts protective ants and acts as a food source for other aphids. This sugary substance encourages aphids to cluster together, enhancing their collective defenses against predators.

Colony dynamics: Cooperation and survival

Aphid colonies operate on a finely tuned balance of cooperation and survival. The presence of mutualistic relationships with ants is a prime example. Ants protect aphids from predators and parasites in exchange for honeydew.

This mutualism benefits both parties: the ants get a reliable food source, and the aphids gain a formidable defense force.

Imagine a bustling aphid city where the ants are the law enforcement, ensuring peace and order in exchange for a sugary tax. This relationship is so beneficial that some aphids have evolved to depend entirely on ants for protection.

The Ant-Aphid Mutualism

In many aphid colonies, ants act as bodyguards. They ward off predators and even move aphids to better feeding sites. This relationship is so beneficial that some aphids have evolved to depend entirely on ants for protection.

Ants are not just bodyguards; they are also aphid herders. They can transport aphids to new plants when the current one is depleted, ensuring a constant supply of honeydew. It's like having your own personal Uber service but with more legs and less tipping.

This mutualistic relationship goes even further: some ant species will carry aphid eggs to their nests during winter, providing a safe environment for the eggs to hatch in the spring. This ensures a ready supply of aphids (and honeydew) when the weather warms up.

The dynamic between ants and aphids is an example of mutualism, where both species benefit from the relationship. Ants derive sustenance from the honeydew produced by aphids, while aphids receive protection from predators. This mutual dependence can shape the behavior and distribution of both species in the ecosystem.

Ants are known to actively farm aphids, manipulating their behavior and movement to maximize honeydew production. This farming behavior can include cutting off the wings of aphids to prevent them from dispersing or herding them to optimal feeding sites. The extent of this manipulation highlights the complex interactions that define the ant-aphid mutualism.

The reproductive strategies of aphids

Reproduction in aphids is a complex and fascinating affair. Most aphids reproduce through parthenogenesis, where females give birth to genetically identical daughters without the need for males. This method allows for rapid population growth, as aphids can produce multiple generations in a single season.

But that's not all—aphids also have a backup plan. When environmental conditions become unfavorable, such as the approach of winter, aphids switch to sexual reproduction. This results in the production of eggs that can withstand harsh conditions, ensuring the survival of the next generation.

This dual reproductive strategy is like having both a fire extinguisher and a sprinkler system in place. If one fails, the other kicks in to save the day.

Parthenogenesis is highly advantageous in stable environments where conditions are favorable for rapid population growth. It allows aphids to exploit resources quickly and efficiently, leading to exponential increases in their numbers.

Sexual reproduction, on the other hand, introduces genetic variation into the population. This genetic diversity is crucial for adapting to changing environmental conditions and for evolving resistance to pathogens and predators. By alternating between parthenogenesis and sexual reproduction, aphids can balance the benefits of rapid population growth with the need for genetic diversity.

Case study: The role of winged aphids

When aphid populations become too dense or food becomes scarce, some aphids develop wings and migrate to new plants. These winged aphids are crucial for the spread of aphid populations and the colonization of new habitats. It's like sending out tiny, winged pioneers to find new frontiers.

These winged aphids are the adventurers of their species, boldly going where no aphid has gone before. They ensure the survival and spread of the population, even in the face of adversity.

Winged aphids are not just for spreading out; they play a crucial role in genetic diversity. When they migrate and form new colonies, they often interbreed with aphids from other colonies, introducing new genetic material into the population. This genetic mixing can increase the resilience of aphid populations to environmental changes and pest control measures.

The development of wings in aphids is a phenotypic response to environmental stressors such as crowding or food scarcity. This plasticity in response to environmental cues allows aphids to escape unfavorable conditions and colonize new habitats, ensuring the persistence of the population.

The impact of environmental factors

Environmental factors play a significant role in shaping the social structures and behaviors of aphid colonies. Temperature, humidity, and availability of food resources can influence the size and composition of colonies, as well as the reproductive strategies employed by aphids.

For instance, in warmer climates, aphids might reproduce more rapidly due to favorable conditions, leading to larger colonies. In contrast, in colder regions, aphids may rely more on sexual reproduction to produce hardy eggs that can survive the winter.

Aphids have also adapted to seasonal changes by adjusting their reproductive strategies. During the spring and summer, when conditions are optimal, they reproduce asexually and rapidly increase their numbers. As the seasons change and conditions become less favorable, they switch to sexual reproduction, producing eggs that can withstand the cold winter months.

These seasonal adjustments ensure that aphids can survive and thrive throughout the year, regardless of environmental changes. Their ability to adapt to varying conditions is a testament to their evolutionary success.

Temperature is not the only environmental factor that affects aphid colonies. Rainfall and humidity can also play significant roles. High humidity levels can favor the growth of fungal pathogens that can decimate aphid populations. In contrast, dry conditions can make it easier for aphids to survive and reproduce.

Light availability also influences aphid behavior and reproduction. Aphids are photoperiod sensitive, meaning they respond to the length of day and night. Changes in light conditions can trigger the production of winged forms or the switch from asexual to sexual reproduction.

Understanding how environmental factors influence aphid populations is critical for developing effective pest management strategies. By predicting how aphid populations will respond to changes in temperature, humidity, and other environmental variables, farmers and pest control professionals can implement timely interventions to minimize crop damage.

Aphid defense mechanisms

Aphids are not defenseless, and their survival strategies go beyond just running away. Some species produce waxy filaments that cover their bodies, making them less palatable to predators. Others can secrete sticky substances that trap their attackers.

Aphids are also known to employ chemical defenses. Some species can produce toxic compounds that deter predators. These chemicals can make the aphids taste bad or even be harmful if ingested.

These defense mechanisms are like having a personal security system and a supply of pepper spray. They ensure that aphids can protect themselves from various threats, increasing their chances of survival.

In addition to physical and chemical defenses, aphids also use behavioral strategies to avoid predation. For example, some aphids drop off their host plant when disturbed, effectively escaping from predators. This behavior, known as "dropping," can be a highly effective escape mechanism, especially when combined with their small size and ability to blend into their surroundings.

Aphids can also form symbiotic relationships with other insects, such as ants, which provide additional protection. By integrating multiple defense strategies, aphids can increase their chances of survival in a hostile environment.

The evolution of these diverse defense mechanisms highlights the adaptive flexibility of aphids. By employing a combination of physical, chemical,

and behavioral defenses, aphids can effectively deter a wide range of predators, from small insects to larger vertebrates.

The role of symbiotic bacteria

Aphids harbor symbiotic bacteria within their bodies, which play a crucial
role in their survival. These bacteria help aphids digest plant sap, providing
essential nutrients that aphids cannot obtain on their own. This symbiotic
relationship is so important that aphids have evolved to pass these bacteria
to their offspring.

This relationship is akin to having an in-house chef who also happens to
be a nutritionist, ensuring that every meal is perfectly balanced.

The symbiotic bacteria, known as Buchnera, reside in specialized cells
called bacteriocytes. These bacteria provide aphids with essential amino acids and other nutrients that are lacking in their diet of plant sap. In return,
the bacteria benefit from a stable environment within the aphid's body.

This mutualistic relationship is so integral to aphid survival that it has
shaped their evolutionary history. Aphids and Buchnera have co-evolved
for millions of years, resulting in a highly specialized and interdependent
relationship.

Buchnera bacteria are passed from mother to offspring during reproduction, ensuring that each new generation of aphids inherits these essential
symbionts. This vertical transmission of bacteria is critical for maintaining
the mutualistic relationship and ensuring the survival of both the aphids and
their bacterial partners.

The symbiotic relationship between aphids and Buchnera is a prime example of mutualistic coevolution. The dependence of aphids on Buchnera
for essential nutrients has led to the loss of certain metabolic pathways in
aphids, making them entirely reliant on their bacterial partners. In turn,
Buchnera bacteria have undergone genome reduction, losing many genes
unnecessary for their symbiotic lifestyle but retaining those crucial for nu-
trient synthesis.

Aphids in agricultural ecosystems

In agricultural ecosystems, aphids are often considered pests due to their
feeding habits and the plant viruses they can transmit. However, their interactions with other species, such as their mutualistic relationships with
ants and their role as prey for various predators, highlight their ecological
significance.

Aphids play a crucial role in the food web, serving as a primary food
source for many predators, including ladybirds, lacewings, and parasitoid
wasps. These predators help regulate aphid populations, preventing them
from becoming too numerous and causing significant damage to crops.

Understanding the complex social structures and behaviors of aphids can provide insights into managing their populations more effectively. By disrupting their communication methods or targeting their reproductive strategies, we can develop more sustainable pest control methods.

Integrated pest management (IPM) strategies that incorporate biological control agents, such as natural predators and parasitoids, can help reduce aphid populations without relying heavily on chemical pesticides. These strategies not only protect crops but also promote a healthier and more balanced ecosystem.

The use of biological control agents, such as ladybird beetles and parasitoid wasps, has been shown to be effective in reducing aphid populations. These natural enemies can significantly impact aphid numbers, leading to improved crop health and reduced reliance on chemical pesticides.

Biological control agents can be introduced into agricultural systems in various ways. For instance, releasing ladybird beetles (commonly known as ladybugs) into infested fields can help reduce aphid populations. Similarly, parasitoid wasps can be introduced to target specific aphid species, laying their eggs inside the aphids and ultimately killing them. Gruesome, but effective!

The implementation of IPM strategies requires careful monitoring of aphid populations and their natural enemies. By maintaining a balance between aphids and their predators, farmers can minimize crop damage while reducing the need for chemical interventions.

The importance of studying aphid behavior

Studying aphid behavior and social structures is not just an academic exercise—it has practical applications in agriculture and pest management. By understanding how aphids communicate, reproduce, and interact with their environment, we can develop targeted strategies to control their populations and minimize their impact on crops.

For example, researchers have explored the use of pheromone traps to monitor and manage aphid populations. These traps attract aphids using synthetic pheromones, allowing farmers to detect and respond to infestations early. This proactive approach can help prevent aphid outbreaks and reduce the need for chemical interventions.

Additionally, understanding the mutualistic relationships between aphids and ants can inform strategies to disrupt these interactions. By preventing ants from protecting aphid colonies, we can make aphids more vulnerable to natural predators and reduce their numbers.

Aphid behavior research can also lead to the development of novel pest control methods. For instance, scientists are investigating the use of RNA interference (RNAi) to disrupt key genes involved in aphid reproduction and survival. By targeting these genes, it may be possible to reduce aphid populations in a species-specific manner, minimizing the impact on non-target organisms.

RNAi technology involves introducing double-stranded RNA molecules into aphid populations, which can specifically target and silence essential genes. This approach has the potential to provide a highly targeted and environmentally friendly method of pest control, reducing the reliance on broad-spectrum chemical pesticides.

Understanding aphid behavior can also inform the development of resistant crop varieties. By identifying the genetic basis of aphid resistance in certain plants, scientists can breed or genetically engineer crops that are less susceptible to aphid feeding and the viruses they transmit.

Future directions in aphid research

As our understanding of aphid biology and behavior continues to grow, new research avenues and technologies are emerging. Advances in molecular biology and genomics are providing insights into the genetic basis of aphid traits, such as their reproductive strategies and resistance to pesticides.

Genomic studies have revealed the genetic diversity within aphid populations and identified genes associated with key traits, such as host plant adaptation and insecticide resistance. These findings can inform the development of more targeted and effective pest control measures.

Furthermore, researchers are exploring the potential of biocontrol agents, such as entomopathogenic fungi and nematodes, to manage aphid populations. These natural enemies can infect and kill aphids, offering a sustainable and environmentally friendly alternative to chemical pesticides.

The integration of these advanced technologies and approaches into IPM programs holds promise for more effective and sustainable aphid management. By combining traditional ecological knowledge with cutting-edge research, we can develop holistic strategies that protect crops and preserve the health of agricultural ecosystems.

Researchers are also investigating the potential impact of climate change on aphid populations and their interactions with other species. Changes in temperature and precipitation patterns can influence aphid reproduction, migration, and the dynamics of their mutualistic and antagonistic relationships. Understanding these impacts is crucial for developing adaptive pest management strategies in a changing climate.

Another exciting area of research is the use of genetic engineering to enhance the natural resistance of crops to aphids. By introducing genes that confer resistance to aphid feeding or the viruses they transmit, scientists aim to develop crop varieties that are less susceptible to aphid infestations, reducing the need for chemical pesticides.

Genetic engineering approaches can include the introduction of genes that produce antifeedant compounds or enhance the plant's natural defense mechanisms. These genetically modified crops can provide a sustainable solution to aphid infestations, reducing the economic and environmental costs associated with traditional pest control methods.

Conclusion

The social structure, communication methods, and colony dynamics of aphids reveal a world of complexity and cooperation that rivals any human society. These tiny insects, often seen as mere pests, have developed sophisticated strategies to thrive in diverse environments. Understanding the intricacies of aphid life not only highlights their ecological importance but also offers insights into managing their populations in agricultural settings.

As we continue to explore the world of aphids, it's clear that these tiny creatures have much to teach us about survival, cooperation, and the delicate balance of nature. By studying their behaviors and interactions, we can develop more effective and sustainable methods for managing aphid populations and protecting our crops.

Glossary of Terms

Alarm Pheromones: Chemical signals released by aphids when attacked, causing others to flee.

Aphidoidea: A superfamily of insects in the order Hemiptera, commonly known as aphids.

Buchnera: Symbiotic bacteria that reside within aphids and provide essential nutrients.

Colony Dynamics: The interactions and organizational structure within a colony.

Entomopathogenic Fungi: Fungi that infect and kill insects, used as biological control agents.

Honeydew: A sugary substance excreted by aphids that attracts ants.

Integrated Pest Management (IPM): An approach to pest control that combines biological, cultural, and chemical methods to manage pest populations effectively.

Mutualism: A symbiotic relationship where both parties benefit.

Nematodes: Microscopic roundworms that can infect and kill insects, used as biological control agents

Parthenogenesis: A form of asexual reproduction where an organism can produce offspring without mating.

Soldier Aphids: Specialized aphids that defend the colony against predators.

Viviparity: The condition where organisms give birth to live young, bypassing the egg stage.

Winged Aphids: Aphids that develop wings to migrate to new plants when conditions become unfavorable.

Bibliography

Blackman, R. L., & Eastop, V. F. (2000). *Aphids on the World's Crops: An Identification and Information Guide.* Wiley.

Dixon, A. F. G. (1998). *Aphid Ecology: An Optimization Approach.* Chapman & Hall.

Douglas, A. E. (1998). Nutritional interactions in insect-microbial symbioses: Aphids and their symbiotic bacteria Buchnera. *Annual Review of Entomology, 43*(1), 17–37.

Hogenhout, S. A., & Bos, J. I. (2011). Effector proteins that modulate plant–insect interactions. *Current Opinion in Plant Biology, 14*(4), 422–428.

Moran, N. A., & Baumann, P. (2000). Bacterial endosymbionts in animals. *Current Opinion in Microbiology, 3*(3), 270–275.

Van Emden, H. F., & Harrington, R. (2007). *Aphids as Crop Pests.* CABI.

The aphid arsenal

Introduction

Picture an aphid, that tiny sap-sucking insect you've probably flicked off a plant without a second thought. Now imagine that same aphid outfitted with an arsenal of survival tools that would make a secret agent jealous. Aphids are not just defenseless, squishy bugs; they are masters of mimicry, chemical warfare, and strategic alliances. Welcome to the world of the aphid arsenal, where tiny warriors use an impressive array of defenses to outwit and outlive their predators.

The art of mimicry: Blending in to stay safe

One of the most effective ways aphids avoid becoming a snack is by blending in with their surroundings. Mimicry, the ability to resemble another organism or part of the environment, is a common defense mechanism among aphids. This camouflage helps them go unnoticed by predators.

Aphids can take on the color of the plant they inhabit, making them virtually invisible. This adaptive coloration is not just limited to green; aphids can appear yellow, red, or even black, depending on their host plant. Some species go a step further, mimicking plant parts like thorns or galls to avoid detection.

Case study: The pea aphid (*Acyrthosiphon pisum*)

The pea aphid is a master of disguise. Depending on the host plant, these aphids can change color to blend in perfectly. When on alfalfa, they turn green; on red clover, they turn pink. This color change is controlled by the plant's chemical signals, which the aphids detect and respond to. It's like having a built-in wardrobe that changes with the seasons!

DOI: 10.1201/9781003537786-3

> **Aphid Mimicry Techniques**
> - **Color Adaptation:** Changing color to match the host plant.
> - **Mimicking Plant Parts:** Resembling thorns or galls to avoid detection.
> - **Chemical Detection:** Sensing and responding to plant chemical signals for camouflage.

Chemical warfare: Aphids' toxic arsenal

Aphids have developed a range of chemical defenses to protect themselves from predators. These defenses can include the production of toxic compounds, alarm pheromones, and sticky secretions.

Alarm pheromones

When an aphid is attacked, it releases alarm pheromones, primarily composed of (E)-β-farnesene. This chemical signal prompts nearby aphids to disperse quickly, reducing the chance of mass casualties. It's the aphid equivalent of yelling "Fire!" in a crowded theater.

Sticky secretions

Some aphids can produce a sticky substance that traps predators. This gooey defense can deter small predators like ants and spiders. It's akin to setting up a glue trap for your enemies—simple but effective.

> **Aphid Chemical Defenses**
> - **Alarm Pheromones:** Chemical signals that prompt other aphids to flee.
> - **Toxic Compounds:** Substances that make aphids unpalatable or harmful to predators.
> - **Sticky Secretions:** Glue-like substances that trap predators.

These chemical defenses are remarkably effective. For example, the green peach aphid (*Myzus persicae*) releases alarm pheromones that can cause entire colonies to scatter, while the sticky secretions of the woolly aphid (*Eriosoma lanigerum*) can trap and immobilize predators.

Alliance with ants: The ultimate bodyguards

Aphids have formed mutualistic relationships with ants, which act as their bodyguards in exchange for honeydew, a sugary substance aphids excrete. This alliance provides aphids with protection from predators and parasites.

Ants diligently tend to their aphid "cows," moving them to better feeding sites, protecting them from threats, and even carrying them to safety when danger looms. In return, ants harvest the honeydew produced by aphids, which serves as a valuable food source.

Case study: The black bean aphid (*Aphis fabae*) and ants

The black bean aphid has a symbiotic relationship with various ant species. Ants protect these aphids from ladybirds and parasitic wasps, ensuring their survival. This relationship is so beneficial that aphids have evolved to produce more honeydew, enhancing their appeal to ants. It's a win-win situation: the ants get a steady supply of food, and the aphids get a personal security detail.

The Ant-Aphid Mutualism

- **Protection:** Ants defend aphids from predators.
- **Transportation:** Ants move aphids to optimal feeding locations.
- **Nutrition:** Aphids provide ants with honeydew.

The mutualistic relationship between ants and aphids is a fascinating example of interspecies cooperation. Ants have even been observed milking aphids for honeydew, gently stroking them with their antennae to stimulate production. This mutual dependence enhances the survival chances of both species.

Figure 3.1 Black bean aphid. (Shutterstock ID: 2394960067)

Mechanical defenses: The power of waxy coatings

Many aphids produce a waxy coating that covers their bodies, making them less palatable to predators and difficult to grasp. This wax can also protect against desiccation and environmental stress.

The waxy coating is often produced by specialized glands and can vary in thickness and composition. Some aphids even produce long, filamentous wax strands that can entangle small predators.

Case study: The woolly apple aphid (*Eriosoma lanigerum*)

The woolly apple aphid is covered in a dense, wool-like wax that provides protection against predators and parasitoids. This wax also helps the aphid regulate moisture and temperature, allowing it to survive in various climates. It's like having a built-in suit of armor and a climate control system all in one!

Mechanical Defenses

- **Waxy Coatings:** Protective layers that deter predators.
- **Filamentous Wax Strands:** Long strands that can entangle predators.
- **Moisture Regulation:** Wax helps prevent desiccation and regulate temperature.

The effectiveness of waxy coatings is evident in the reduced predation rates observed in wax-covered aphids. This mechanical defense is particularly useful in harsh environments where aphids must contend with both predators and environmental stressors.

Evasive maneuvers: Dropping and rolling

When threatened, some aphids employ evasive maneuvers such as dropping off the plant to escape predators. This behavior, known as "dropping," is a rapid response to physical disturbances.

Once on the ground, aphids can use their legs to roll away or find a new plant to climb. This quick escape mechanism reduces the likelihood of being caught by predators that specialize in ambush tactics.

Evasive Maneuvers

- **Dropping:** Falling off the plant to escape predators.
- **Rolling:** Using legs to roll away from danger.
- **Climbing:** Finding new plants to avoid returning to the same threat.

These evasive maneuvers are surprisingly effective. Studies have shown that dropping can significantly reduce predation risk, especially from predators like ladybird beetle larvae that rely on surprise attacks. By employing such rapid escape tactics, aphids can avoid becoming an easy meal.

Symbiosis with endosymbionts: Microbial allies

Aphids rely on endosymbiotic bacteria, particularly *Buchnera aphidicola,* to provide essential nutrients that are lacking in their diet of plant sap. These bacteria reside within specialized cells called bacteriocytes and produce amino acids and vitamins crucial for aphid survival.

This symbiotic relationship is so integral to aphid biology that *Buchnera* is passed maternally from one generation to the next. Without these microbial allies, aphids would be unable to thrive on their nutritionally poor diet.

Case study: The pea aphid (Acyrthosiphon pisum) and *Buchnera*

The pea aphid's relationship with *Buchnera* is a classic example of mutualistic symbiosis. *Buchnera* provide essential amino acids, allowing the aphids to thrive on plant sap. In return, *Buchnera* benefits from a stable environment within the aphid's body. This relationship has evolved over millions of years, resulting in highly specialized and interdependent partners.

Sidebar: Endosymbiotic Relationships

- *Buchnera aphidicola*: Provides essential nutrients to aphids.
- **Bacteriocytes:** Specialized cells housing endosymbionts.
- **Nutritional Mutualism:** Both partners benefit from the exchange of nutrients.

The co-evolution of aphids and *Buchnera* highlights the deep interdependence of their relationship. The genetic integration between aphids and their endosymbionts is so profound that *Buchnera* cannot survive outside the aphid host, and aphids cannot thrive without *Buchnera*.

Adaptive reproduction: A two-pronged strategy

Aphids exhibit a remarkable reproductive strategy that includes both asexual and sexual reproduction. This dual approach allows them to rapidly increase their populations while also ensuring genetic diversity to adapt to changing environmental conditions.

During favorable conditions, aphids reproduce asexually through parthenogenesis, producing clones of themselves. When conditions become harsh, they switch to sexual reproduction, producing eggs that can withstand adverse environments.

Case study: The green peach aphid (*Myzus persicae*)

The green peach aphid exemplifies adaptive reproduction. In the spring and summer, they reproduce asexually, creating large colonies quickly. As winter approaches, they switch to sexual reproduction, laying eggs that can survive the cold. This flexible reproductive strategy allows them to exploit favorable conditions and endure challenging ones.

Reproductive Strategies
- **Parthenogenesis:** Asexual reproduction producing clones.
- **Sexual Reproduction:** Producing eggs for genetic diversity.
- **Seasonal Adaptation:** Switching strategies based on environmental conditions.

This dual reproductive strategy ensures that aphids can rapidly colonize new environments while maintaining the genetic diversity needed to adapt to changing conditions. It's a perfect balance between exploiting the present and preparing for the future.

The role of environmental factors in defense

Environmental factors such as temperature, humidity, and light can influence aphid defense mechanisms. For instance, aphids may alter their wax production or reproductive strategies in response to environmental changes.

Temperature can affect the efficacy of aphid defenses. In colder climates, aphids may produce thicker wax coatings to prevent desiccation and insulate against the cold. Conversely, in warmer climates, they may produce more honeydew to attract ants for protection.

Case study: The Russian wheat aphid (*Diuraphis noxia*)

The Russian wheat aphid adjusts its defenses based on environmental conditions. In colder regions, it produces a thicker wax coating to survive the harsh climate. In warmer areas, it relies more on mutualistic relationships with ants to protect against predators. This adaptability ensures its survival across diverse environments.

Environmental Adaptations
- **Wax Production:** Adjusting thickness based on temperature.
- **Honeydew Production:** Varying amounts to attract ant protection.
- **Reproductive Shifts:** Changing strategies with seasons.

Understanding how environmental factors influence aphid defenses can help in developing more effective pest management strategies. By predicting how aphid populations will respond to changes in temperature, humidity, and other environmental variables, farmers and pest control professionals can implement timely interventions to minimize crop damage.

Future directions in aphid defense research

As our understanding of aphid defense mechanisms continues to grow, new research avenues and technologies are emerging. Advances in molecular biology and genomics are providing insights into the genetic basis of aphid traits, such as their defensive strategies and resistance to pesticides.

Researchers are exploring the potential of genetic engineering to enhance the natural defenses of crops against aphids. By introducing genes that produce antifeedant compounds or enhance the plant's natural defense mechanisms, scientists aim to develop crop varieties that are less susceptible to aphid infestations.

The integration of these advanced technologies into pest management programs holds promise for more effective and sustainable aphid control. By combining traditional ecological knowledge with cutting-edge research, we can develop holistic strategies that protect crops and preserve the health of agricultural ecosystems.

Conclusion

Aphids, often seen as mere garden pests, possess an impressive arsenal of defense mechanisms and survival strategies. From mimicry and chemical warfare to alliances with ants and symbiotic relationships with bacteria, aphids demonstrate a remarkable ability to adapt and thrive in their environments.

Understanding these defenses not only highlights the ecological significance of aphids but also offers insights into developing more effective and sustainable pest management strategies. As we continue to explore the fascinating world of aphid defenses, it becomes clear that these tiny insects have much to teach us about survival, resilience, and the complex interplay of nature's defense mechanisms.

Glossary of Terms

Alarm Pheromones: Chemical signals released by aphids when attacked, causing others to flee.

Antifeedant Compounds: Chemicals that deter herbivores from feeding on plants.

Bacteriocytes: Specialized cells in aphids that house symbiotic bacteria.

Buchnera aphidicola: Symbiotic bacteria that reside within aphids and provide essential nutrients.

Entomopathogenic Fungi: Fungi that infect and kill insects, used as biological control agents.

Honeydew: A sugary substance excreted by aphids that attracts ants.

Mimicry: The ability of an organism to resemble another organism or part of the environment to avoid detection by predators.

Mutualism: A symbiotic relationship where both parties benefit.

Parthenogenesis: A form of asexual reproduction where an organism can produce offspring without mating.

Symbiosis: A close and often long-term interaction between two different biological species.

Bibliography

Blackman, R. L., & Eastop, V. F. (2000). *Aphids on the World's Crops: An Identification and Information Guide*. Wiley.

Dixon, A. F. G. (1998). *Aphid Ecology: An Optimization Approach*. Chapman & Hall.

Douglas, A. E. (1998). Nutritional interactions in insect-microbial symbioses: Aphids and their symbiotic bacteria Buchnera. *Annual Review of Entomology, 43*(1), 17–37.

Hogenhout, S. A., & Bos, J. I. (2011). Effector proteins that modulate plant–insect interactions. *Current Opinion in Plant Biology, 14*(4), 422–428.

Stadler, B., & Dixon, A. F. G. (2005). Ecology and evolution of aphid-ant interactions. *Annual Review of Ecology, Evolution, and Systematics, 36*, 345–372.

Van Emden, H. F., & Harrington, R. (2007). *Aphids as Crop Pests*. CABI.

CHAPTER 4

Love bugs: Aphid romance revealed

Introduction

Aphids might not be the first insects that come to mind when you think of romance, but their love lives are surprisingly intricate and fascinating. These tiny insects engage in a range of reproductive strategies that allow them to thrive in diverse environments and rapidly colonize new habitats. Welcome to the world of aphid romance, where mating rituals, reproductive strategies, and lifecycle dynamics come together to create a true love bug saga.

When we think of aphids, most of us picture a tiny, sap-sucking pest causing havoc in our gardens. Rarely do we consider the dramatic, behind-the-scenes world of aphid reproduction—a world that rivals the most complex soap operas. Aphids have evolved a multitude of strategies to ensure their survival and proliferation, from cloning themselves in an impressive feat of asexual reproduction to engaging in surprisingly romantic (for insects, anyway) mating dances when the situation calls for it.

The reproductive strategies of aphids are not just about producing the next generation; they are intricately tied to the aphids' ability to adapt to their environment. Whether it's rapidly increasing their numbers to take advantage of a bountiful season or mixing their genetic material to weather harsh conditions, aphids are the ultimate survivors. Their lifecycles, influenced by factors like temperature, food availability, and daylight, are a testament to nature's ingenuity.

In this chapter, we will delve into the lifecycle of aphids, starting from their beginnings as eggs and following their journey through nymphhood to adulthood. We'll explore the different reproductive strategies they employ—both

DOI: 10.1201/9781003537786-4

asexual and sexual—and understand how these methods contribute to their rapid population growth.

Join us as we venture into the hidden world of aphid romance, where tiny insects engage in epic reproductive battles and intimate mating rituals. By the end of this chapter, you'll see aphids in a whole new light—not just as garden pests, but as master strategists of the insect world, employing a complex web of reproductive tactics to ensure their survival.

So, grab your magnifying glass and let's dive into the love lives of aphids, where the drama is small in size but immense in impact.

The lifecycle of the aphid: From egg to adult

Understanding the aphid lifecycle is key to appreciating their reproductive strategies. The lifecycle of an aphid can be broadly divided into several stages: egg, nymph, and adult. Each stage plays a critical role in the survival and reproduction of these insects.

Egg stage

The aphid lifecycle often begins with an egg. During harsh conditions, such as winter, many aphid species switch to sexual reproduction and produce eggs that can withstand adverse environments. These eggs are typically laid on the bark or leaves of host plants and remain dormant until favorable conditions return.

Case study: The green peach aphid (*Myzus persicae*)

The green peach aphid lays its eggs on peach trees in the fall. These eggs, known as overwintering eggs, are highly resistant to cold temperatures and can survive the winter months. When spring arrives, the eggs hatch into nymphs, ready to start the cycle anew.

Aphid Egg Characteristics
- **Overwintering Eggs:** Resistant to cold temperatures.
- **Egg Laying Sites:** Often on bark or leaves of host plants.
- **Dormancy:** Eggs remain dormant until favorable conditions return.

Nymph stage

Upon hatching, aphid eggs release nymphs, which are essentially miniature versions of the adult aphid. Nymphs go through several molts, shedding their exoskeletons as they grow. Each molt brings them closer to adulthood.

Case study: The pea aphid (*Acyrthosiphon pisum*)

The pea aphid nymph goes through four to five molts before reaching adulthood. During each molt, the nymph sheds its old exoskeleton and develops a new one. This process allows the nymph to grow and prepare for the reproductive phase of its lifecycle.

> ### Nymph Development
> - **Molting:** Shedding exoskeleton to grow.
> - **Instars:** Stages between molts.
> - **Rapid Growth:** Nymphs quickly progress to adulthood.

Adult stage

The adult stage is where aphids truly shine in their reproductive prowess. Depending on the environmental conditions, adult aphids can reproduce sexually or asexually. The ability to switch between these reproductive modes allows aphids to maximize their population growth and adapt to changing environments.

Asexual reproduction: Cloning galore

Aphids are well known for their ability to reproduce asexually through a process called parthenogenesis. In this mode, female aphids give birth to live young without the need for males. This method allows for rapid population growth, as each female can produce numerous offspring in a short period.

Case study: The cotton aphid (*Aphis gossypii*)

The cotton aphid reproduces asexually during the growing season, producing multiple generations in a single year. Each female can give birth to dozens of nymphs, which themselves can start reproducing within a week. This exponential growth leads to large colonies capable of infesting crops quickly.

> ### Asexual Reproduction Benefits
> - **Rapid Population Growth:** Quick increase in numbers.
> - **No Mating Required:** Females reproduce without males.
> - **High Offspring Rate:** Multiple generations in a short time.

Sexual reproduction: The aphid mating dance

While asexual reproduction dominates during favorable conditions, aphids switch to sexual reproduction when the environment becomes less hospitable. This typically occurs in the fall when temperatures drop, and food

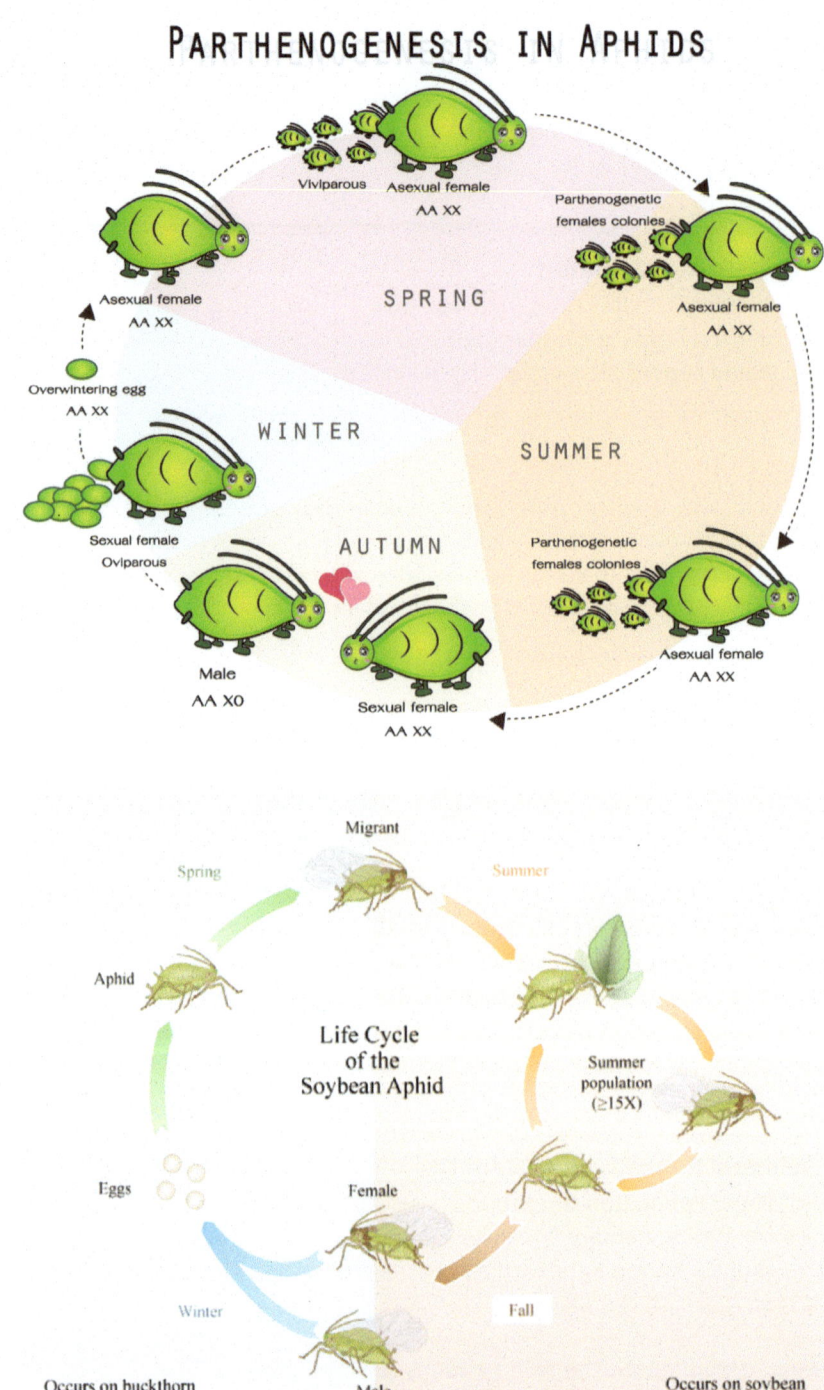

Figure 4.1 Parthenogenesis. (Shutterstock ID: 2004380153)

sources dwindle. Sexual reproduction allows for genetic diversity, which can help aphids survive adverse conditions.

Case study: The rosy apple aphid (*Dysaphis plantaginea*)

The rosy apple aphid engages in sexual reproduction as winter approaches. Male and female aphids are produced, and after mating, females lay overwintering eggs. This genetic mixing provides the population with the diversity needed to adapt to changing environments.

Sexual Reproduction Process
- **Male and Female Production:** Both sexes are produced.
- **Mating:** Occurs before winter.
- **Egg Laying:** Overwintering eggs are laid for the next generation.

Sexual Reproduction Benefits
- **Genetic Diversity:** Increases adaptability.
- **Environmental Adaptation:** Prepares for adverse conditions.
- **Survival of the Fittest:** Ensures the strongest genes are passed on.

Parthenogenesis: The miracle of cloning

Parthenogenesis is the most fascinating aspect of aphid reproduction. During this process, females produce genetically identical daughters without fertilization. This method is advantageous in stable environments where rapid population growth is essential.

Case study: The aphid of the cabbage (*Brevicoryne brassicae*)

The cabbage aphid can produce up to 10 generations per growing season through parthenogenesis. Each generation consists of females that clone themselves, leading to a uniform and rapidly growing population. This reproductive strategy is particularly effective in monoculture crops where environmental conditions remain stable.

Parthenogenesis Facts
- **Cloning:** Produces genetically identical offspring.
- **High Reproductive Rate:** Multiple generations in one season.
- **Uniformity:** Offspring are genetically the same.

The role of environmental factors in reproduction

Environmental factors play a significant role in determining the reproductive strategy aphids use. Temperature, availability of food, and photoperiod (length of day and night) can influence whether aphids reproduce sexually or asexually.

Case study: The black bean aphid (*Aphis fabae*)

The black bean aphid switches between reproductive modes based on environmental cues. During the summer, when food is abundant and temperatures are warm, they reproduce asexually. As fall approaches and conditions become less favorable, they switch to sexual reproduction to produce overwintering eggs.

Environmental Cues
- **Temperature:** Influences reproductive mode.
- **Food Availability:** Abundance favors asexual reproduction.
- **Photoperiod:** Changes trigger reproductive shifts.

Population growth: The power of numbers

Aphids are capable of extraordinary population growth due to their reproductive strategies. A single female aphid can produce a large number of offspring, and if conditions are favorable, these offspring can reproduce within days. This rapid generation turnover leads to exponential population growth.

Case study: The green peach aphid (*Myzus persicae*)

The green peach aphid exemplifies rapid population growth. In favorable conditions, one aphid can give rise to thousands of offspring within a few weeks. This explosive growth can lead to significant infestations on crops, making them a formidable pest.

Population Growth Factors
- **High Reproductive Rate:** Many offspring per female.
- **Short Generation Time:** Quick turnover of generations.
- **Favorable Conditions:** Leads to exponential growth.

The impact of aphid reproduction on agriculture

The reproductive strategies of aphids have significant implications for agriculture. Rapid population growth can lead to severe infestations, causing damage to crops and reducing yields. Understanding the lifecycle and reproductive behavior of aphids is crucial for developing effective pest management strategies.

Case study: The pea aphid (*Acyrthosiphon pisum*) in agriculture

The pea aphid is a major pest in legume crops. Its ability to reproduce rapidly through parthenogenesis allows it to quickly colonize fields. Effective management requires monitoring aphid populations and implementing control measures before infestations reach damaging levels.

Agricultural Impact
- **Crop Damage:** Aphids can cause significant harm to plants.
- **Yield Reduction:** Infestations lower crop yields.
- **Pest Management:** Understanding reproduction aids in control.

Future directions in aphid reproduction research

Research into aphid reproduction continues to uncover new insights and potential strategies for managing these pests. Advances in molecular biology and genetics are providing a deeper understanding of the mechanisms behind aphid reproductive strategies and lifecycle dynamics.

Case study: Genetic research on aphid reproduction

Recent studies have identified key genes involved in parthenogenesis and sexual reproduction in aphids. Manipulating these genes could provide new methods for controlling aphid populations. For example, disrupting the genes that regulate asexual reproduction could reduce the rapid population growth of aphids.

Research Highlights
- **Genetic Mechanisms:** Understanding the genes behind reproduction.
- **Molecular Tools:** New methods for controlling aphid populations.
- **Future Applications:** Potential for innovative pest management strategies.

Online resources: Infographics on aphid reproduction cycles

For a visual exploration of the aphid lifecycle and reproductive strategies, visit our online resource hub. You'll find detailed infographics and interactive content that illustrate the stages of the aphid lifecycle, the differences between asexual and sexual reproduction, and the impact of environmental factors on aphid reproduction.

Conclusion

The reproductive strategies and lifecycle of aphids reveal a world of complexity and adaptability. From parthenogenesis to sexual reproduction, aphids employ a range of tactics to ensure their survival and rapid population growth. Understanding these strategies not only highlights the fascinating biology of aphids but also offers insights into developing more effective pest management techniques. As we continue to study aphid reproduction, we gain valuable knowledge that can help us better protect crops and maintain healthy agricultural ecosystems.

Glossary of Terms

Asexual Reproduction: A mode of reproduction where offspring are produced by a single parent without the involvement of gamete fusion.

Bacteriocytes: Specialized cells in aphids that house symbiotic bacteria.

Buchnera aphidicola: Symbiotic bacteria that reside within aphids and provide essential nutrients.

Instars: The stages between molts in the lifecycle of an insect.

Molting: The process of shedding the exoskeleton to allow for growth in insects.

Nymph: An immature stage of an insect that undergoes gradual metamorphosis before becoming an adult.

Overwintering Eggs: Eggs laid by aphids that can survive harsh winter conditions and hatch in the spring.

Parthenogenesis: A form of asexual reproduction where an organism produces offspring without fertilization.

Photoperiod: The length of day and night, which can influence biological processes in organisms.

Sexual Reproduction: A mode of reproduction involving the fusion of male and female gametes to create genetic diversity.

Bibliography

Blackman, R. L., & Eastop, V. F. (2000). *Aphids on the World's Crops: An Identification and Information Guide*. Wiley.

Dixon, A. F. G. (1998). *Aphid Ecology: An Optimization Approach*. Chapman & Hall.

Moran, N. A., & Dunbar, H. E. (2006). Sexual acquisition of beneficial symbionts in aphids. *Proceedings of the National Academy of Sciences, 103*(32), 12803–12806.

Simon, J. C., Rispe, C., & Sunnucks, P. (2002). Ecology and evolution of sex in aphids. *Trends in Ecology & Evolution, 17*(1), 34–39.

Van Emden, H. F., & Harrington, R. (2007). *Aphids as Crop Pests*. CABI.

Aphids and agriculture: A rocky relationship

Introduction

When it comes to agriculture, aphids are the equivalent of that annoying neighbor who just won't go away. They're small, they're persistent, and they seem to have an insatiable appetite for your prized plants. Aphids and agriculture have a long, tumultuous relationship, characterized by economic losses, crop damage, and endless battles between farmers and these tiny pests.

Imagine a peaceful farm, crops swaying gently in the breeze, when suddenly—bam!—a horde of aphids descends, turning lush fields into a battleground. These little sap suckers, barely visible to the naked eye, have been a scourge for farmers for centuries. They don't just nibble on plants; they form colonies, reproduce at alarming rates, and can transmit plant viruses that devastate entire crops. The economic ramifications are significant, with aphid infestations causing billions of dollars in losses globally each year.

In this chapter, we will explore this rocky relationship in detail, shedding light on the economic impact of aphids, the myriad challenges in controlling them, and the various methods—both old and new—that have been employed to keep aphid populations under control. From the chemical control methods that aphids have outsmarted to the innovative sustainable practices gaining traction, we'll delve into the ongoing battle against these tiny invaders.

Farmers' frustrations are palpable as they contend with aphids' rapid reproduction and adaptability. Aphids can clone themselves through a process known as parthenogenesis, leading to exponential population growth. Add to this their ability to develop resistance to pesticides, and it's no wonder that farmers often feel like they're fighting a losing battle.

DOI: 10.1201/9781003537786-5

But it's not all doom and gloom. Advances in biological control, genetic research, and integrated pest management (IPM) offer new hope. By combining traditional knowledge with cutting-edge science, researchers and farmers are developing more effective and sustainable ways to manage aphid populations.

Join us as we journey through the history of aphid infestations, examine the latest research in pest control, and explore the stories of farmers on the front lines. Whether you're a farmer, a researcher, or just someone curious about the agricultural challenges posed by aphids, this chapter promises to be both informative and entertaining.

Welcome to the tumultuous world of aphids and agriculture—where every leaf can become a battlefield, and the smallest warriors can have the biggest impacts.

The economic impact of aphids on agriculture

Aphids are not just a minor inconvenience; they are a major economic burden. These tiny insects can cause significant damage to crops, leading to reduced yields and, consequently, substantial financial losses for farmers. The economic impact of aphid infestations is felt worldwide, with certain crops being particularly vulnerable.

> ## Economic Losses Due to Aphid Infestations
> - **Soybeans:** Aphid infestations can lead to yield losses of up to 50%.
> - **Wheat:** Damage from aphid feeding can reduce yields by 10–20%.
> - **Citrus:** Aphids can transmit diseases like the citrus tristeza virus, leading to significant crop losses.
> - **Potatoes:** Aphids are vectors for potato virus Y and leafroll virus, causing up to 80% yield loss.

Case study: The soybean aphid invasion

The soybean aphid (*Aphis glycines*) is a prime example of how aphids can wreak havoc on agriculture. First detected in the United States in 2000, this pest rapidly spread across soybean-growing regions, causing millions of dollars in damage. Farmers had to quickly adapt their pest management strategies to combat this new threat, leading to increased costs and the adoption of more aggressive control measures.

The challenges of aphid control

Controlling aphids is no easy task. These insects are prolific breeders, with some species capable of producing dozens of offspring in just a few days. Their rapid reproduction rates, combined with their ability to develop resistance to pesticides, make them particularly challenging to manage.

Chemical control: The double-edged sword

For many years, chemical pesticides were the go-to solution for controlling aphid populations. However, over time, aphids have developed resistance to many of these chemicals, rendering them less effective. This has led to a cycle of increased pesticide use, higher costs, and growing concerns about environmental and health impacts.

Chemicals That No Longer Work on Aphids

- **Organophosphates:** Once effective, many aphid populations have developed resistance.
- **Pyrethroids:** Resistance is widespread, reducing their efficacy.
- **Neonicotinoids:** Concerns about environmental impact, potential harm to beneficial insects, and resistance have limited their use.

Sustainable Pest Control Methods

With chemical resistance on the rise, farmers and researchers have turned to more sustainable methods of pest control. These approaches aim to manage aphid populations while minimizing harm to the environment and human health.

- **Biological Control:** Introducing natural predators, such as ladybirds and parasitoid wasps, to control aphid populations.
- **Cultural Practices:** Crop rotation, intercropping, and maintaining plant diversity to reduce aphid habitat.
- **Mechanical Control:** Using physical barriers, traps, and manual removal to manage aphid infestations.

Aphid-resistant plants: Nature's defense mechanism

One of the most promising strategies for managing aphid populations is the development of aphid-resistant plant varieties. These plants possess natural defenses that deter aphids, reducing the need for chemical interventions.

Case study: Aphid-Resistant wheat varieties

Researchers have developed wheat varieties that are resistant to the Russian wheat aphid (*Diuraphis noxia*). These

Benefits of Aphid-Resistant Plants

- **Reduced Pesticide Use:** Lower reliance on chemical pesticides.
- **Increased Yields:** Higher crop productivity due to reduced aphid damage.
- **Environmental Benefits:** Less chemical runoff and lower impact on non-target species.

Figure 5.1 Yellow sticky trap. (Shutterstock ID: 1436124848)

plants produce compounds that are toxic or repellent to aphids, providing a built-in defense mechanism. The adoption of these resistant varieties has led to significant reductions in aphid populations and associated crop damage.

Online resource: Database of aphid-resistant crop varieties
For farmers looking to adopt aphid-resistant plants, our online resource hub offers a comprehensive database of available varieties. This resource provides detailed information on the resistance mechanisms, efficacy, and availability of different crop varieties.

Biological control: Nature's pest management
Biological control involves the use of natural predators, parasites, and pathogens to manage pest populations. This method harnesses the power of nature to keep aphid populations in check, providing an eco-friendly alternative to chemical pesticides.

Case study: Ladybirds vs. Aphids
Ladybirds, also known as ladybugs, are voracious predators of aphids. Ladybirds are actually beetles, not true bugs, and extremely effective in their hungry teenager larva stage. A single ladybird can consume hundreds of aphids in its lifetime, making them an effective natural control agent. In

many agricultural systems, farmers release ladybirds as a biological control measure, significantly reducing aphid populations.

Key Biological Control Agents
- **Ladybirds (*Coccinellidae*):** Effective predators of aphids.
- **Parasitoid Wasps (*Aphidiinae*):** Lay eggs inside aphids, leading to their death.
- **Lacewings (*Chrysopidae*):** Larvae are aggressive aphid predators.
- **Fungal Pathogens:** Entomopathogenic fungi can infect and kill aphids.

Implementing Biological Control
- **Predator Release:** Introducing natural enemies into the crop environment.
- **Habitat Management:** Creating conditions that support natural predator populations.
- **Monitoring and Assessment:** Regularly evaluating the effectiveness of biological control measures.

Integrated pest management (IPM): A holistic approach
IPM is a comprehensive strategy that combines multiple pest control methods to manage aphid populations sustainably. IPM focuses on prevention, monitoring, and control, utilizing a combination of biological, cultural, mechanical, and chemical methods.

Case study: IPM in soybean production
In soybean production, IPM strategies have been implemented to manage soybean aphid populations. This approach includes regular monitoring of aphid densities, the use of resistant varieties, biological control agents, and targeted pesticide applications only when necessary. The result is a more sustainable and effective pest management system.

Components of IPM
- **Prevention:** Using resistant varieties and cultural practices to reduce pest pressure.
- **Monitoring:** Regularly assessing pest populations and damage levels.
- **Control:** Implementing a combination of biological, mechanical, and chemical control methods as needed.

Benefits of IPM
- **Sustainability:** Reduces reliance on chemical pesticides.
- **Economic Efficiency:** Lowers costs associated with pest control.
- **Environmental Protection:** Minimizes impact on non-target species and ecosystems.

Future directions in aphid control

The battle against aphids is ongoing, and researchers are continually exploring new strategies and technologies to manage these persistent pests. Advances in molecular biology, genomics, and biotechnology hold promise for developing innovative pest control methods.

Case study: Genetic engineering for aphid resistance

Genetic engineering has the potential to revolutionize aphid control. Researchers are working on developing crop varieties that express genes conferring resistance to aphids. These genetically modified plants produce compounds that deter aphids or interfere with their reproductive processes, offering a powerful tool for pest management.

Emerging Technologies in Aphid Control
- **RNA Interference:** Targeting specific genes to disrupt aphid physiology.
- **CRISPR/Cas9:** Editing plant genomes to enhance resistance traits.
- **Microbial Biocontrol:** Using beneficial microbes to protect plants from aphids.

Ethical and Regulatory Considerations
- **Safety:** Assessing the safety of genetically modified crops.
- **Regulations:** Navigating the regulatory landscape for biotechnology.
- **Public Perception:** Addressing concerns and misconceptions about genetic engineering.

Conclusion

The relationship between aphids and agriculture is indeed rocky, characterized by economic challenges, crop damage, and the constant need for effective pest management strategies. By understanding the economic impact of aphids, the challenges of controlling them, and the various methods employed in their management, we can develop more sustainable and effective approaches to protect our crops.

As we continue to explore new technologies and refine existing methods, the goal remains the same: to manage aphid populations in a way that minimizes economic losses, protects the environment, and ensures the sustainability of agricultural systems. Through collaboration, innovation, and a commitment to sustainability, we can navigate the rocky relationship between aphids and agriculture and build a more resilient future.

Glossary of Terms

Aphid-Resistant Plants: Crop varieties that possess natural defenses against aphid infestation.

Biological Control: The use of natural predators, parasites, or pathogens to manage pest populations.

Chemical Pesticides: Substances used to kill or control pests, including insects, weeds, and pathogens.

CRISPR/Cas9: A genome editing tool that allows for precise modifications to DNA.

Cultural Practices: Agricultural techniques, such as crop rotation and intercropping, used to manage pests and improve soil health.

Entomopathogenic Fungi: Fungi that infect and kill insects, used as biological control agents.

Genetic Engineering: The direct manipulation of an organism's genes using biotechnology.

Integrated Pest Management (IPM): A comprehensive pest control strategy that combines multiple methods to manage pest populations sustainably.

Neonicotinoids: A class of insecticides chemically related to nicotine, used to control pests but associated with environmental concerns.

Organophosphates: A group of insecticides that affect the nervous system of pests, many of which have become less effective due to resistance.

Parthenogenesis: A form of asexual reproduction where an organism produces offspring without fertilization.

Parasitoid Wasps: Wasps that lay their eggs inside other insects, with the developing larvae eventually killing the host.

Photoperiod: The length of day and night, which can influence biological processes in organisms.

Pyrethroids: A class of synthetic insecticides modeled after natural pyrethrins, used to control a variety of pests.

RNA Interference (RNAi): A biological process where RNA molecules inhibit gene expression, used in pest management to target specific genes.

Sustainable Agriculture: Farming practices that maintain and improve environmental health, economic profitability, and social equity.

Bibliography

Blackman, R. L., & Eastop, V. F. (2000). *Aphids on the World's Crops: An Identification and Information Guide*. Wiley.

Dedryver, C. A., Le Ralec, A., & Fabre, F. (2010). The conflicting relationships between aphids and men: A review of aphid damage and control strategies. *Comptes Rendus Biologies, 333*(6-7), 539–553.

Dixon, A. F. G. (1998). *Aphid Ecology: An Optimization Approach*. Chapman & Hall.

Makkouk, K. M., & Kumari, S. G. (2009). Viruses infecting pepper crops. *Advances in Virus Research, 75*, 1–35.

Poppy, G. M., & Sutherland, J. P. (2004). Can biological control resolve the aphid problem? In *Advances in Aphid Research* (pp. 241–260). Springer.

Van Emden, H. F., & Harrington, R. (2007). *Aphids as Crop Pests*. CABI.

CHAPTER 6

The gardeners' woe

Introduction

Imagine this: you've spent weeks meticulously tending to your garden, watching with pride as your hibiscus blooms in radiant colors and your vegetable patch thrives under the sun. Then one day, you walk outside, coffee in hand, only to notice tiny, sap-sucking invaders have turned your garden paradise into a battleground. Welcome to the gardener's perennial struggle with aphids.

Aphids are the bane of many a green thumb, turning lush, healthy plants into sticky, wilted shadows of their former selves. These tiny pests come in various colors—green, black, yellow, red, and brown—and can infest virtually any plant. Whether it's the beautiful hibiscus in your flower bed or the tomatoes in your vegetable patch, aphids seem to have an insatiable appetite for your hard-earned gardening success.

Aphids not only damage plants directly by sucking sap, which weakens the plant and distorts its growth, but they also excrete a sticky substance called honeydew. This honeydew attracts ants and promotes the growth of sooty mold, adding insult to injury as your once-pristine garden turns into a sticky, blackened mess. The sight of aphids swarming over your cherished plants can be both disheartening and infuriating, leading many gardeners to feel as though they are fighting a losing battle.

In this chapter, we will explore the personal battles gardeners face against aphids, sharing stories from the trenches and offering practical advice for organic pest control. From the delicate petals of hibiscus to the robust foliage of your cherished food garden, we'll cover the challenges, the solutions, and the stories that make gardening both a joy and a test of patience.

DOI: 10.1201/9781003537786-6

Case study: Hibiscus and the aphid invasion

Hibiscus plants are a favorite target for aphids, particularly the hibiscus aphid (Aphis gossypii). These tiny pests can cause significant damage, leading to distorted leaves, yellowing, and a general decline in plant health. For many gardeners, the sight of a once-vibrant hibiscus covered in aphids is a heartbreaking experience.

Case study: Hibiscus production in the ornamental plant industry

In the ornamental plant industry, hibiscus is a popular plant due to its vibrant flowers and tropical appeal. However, aphid infestations can severely impact the quality and marketability of these plants. A study conducted by the University of Florida's Institute of Food and Agricultural Sciences (UF/IFAS) highlighted the challenges faced by hibiscus growers in managing aphid populations. The study found that hibiscus aphids can reduce plant vigor and flowering, leading to significant economic losses.

> **Signs of Aphid Infestation on Hibiscus**
> - Curling and yellowing leaves
> - Sticky honeydew on leaves and stems
> - Presence of ants around the plant
> - Sooty mold growth

Figure 6.1 Hibiscus. (Shutterstock ID: 2139727969)

Natural control methods for hibiscus aphids

1   **Water Spray:** A strong jet of water can dislodge aphids from hibiscus plants. This method is most effective for small infestations.
2   **Neem Oil:** Neem oil is an organic pesticide that can effectively control aphids. Mix according to label instructions and spray on the affected plants.
3   **Ladybugs:** Introducing ladybugs to your garden can help control aphid populations naturally. These beneficial insects are voracious aphid predators.
4   **Insecticidal Soap:** Insecticidal soaps can be used to smother aphids. Be sure to cover the entire plant, including the undersides of leaves.

The personal food garden: Aphids' buffet

Personal food gardens are another favorite haunt for aphids. From leafy greens to juicy tomatoes, aphids seem to have a particular fondness for the plants we cherish most. This section explores the common challenges faced by gardeners growing their food and offers practical tips for organic pest control.

Figure 6.2 Ladybug attack. (Shutterstock ID: 2077240723)

Case study: Aphids in vegetable gardens (University of California Agriculture and Natural Resources)

The University of California Agriculture and Natural Resources provides extensive resources for managing aphids in vegetable gardens. According to their research, aphids are particularly problematic for crops like tomatoes, peppers, cucumbers, lettuce, and kale. These plants are often targeted by aphids due to their tender new growth and high nutritional content.

> **Common Food Garden Plants Favored by Aphids**
> - Tomatoes
> - Peppers
> - Cucumbers
> - Lettuce
> - Kale

Natural solutions for food garden aphids

1 **Companion Planting:** Planting herbs like basil, dill, or cilantro near your vegetables can help deter aphids. These herbs emit strong scents that aphids dislike.

2 **Garlic Spray:** A homemade garlic spray can repel aphids. Crush a few garlic cloves, steep them in water overnight, and strain the mixture into a spray bottle.

3 **Reflective Mulch:** Reflective mulches, such as aluminum foil, can confuse aphids and reduce their lending rates on plants.

4 **Yellow Sticky Traps:** These traps can catch flying aphids before they settle on your plants. Place them around your garden for effective control.

Flower gardens: Aphid havens

Flower gardens with their vibrant colors and lush foliage can quickly become infested with aphids. These pests are particularly fond of tender new growth and can wreak havoc on a variety of flowering plants.

Case study: Aphid management in ornamental flower gardens (American horticultural society)

The American Horticultural Society (AHS) emphasizes the importance of integrated pest management in ornamental flower gardens. Aphids can cause significant damage to popular flowering plants such as roses, zinnias, nasturtiums, marigolds, and sunflowers. The AHS recommends a combination of cultural, mechanical, and biological control methods to manage aphid populations effectively.

> **Common Flowering Plants Affected by Aphids**
> - Roses
> - Zinnias
> - Nasturtiums
> - Marigolds
> - Sunflowers

Natural aphid control for flower gardens

1 **Beneficial Insects:** Introducing beneficial insects like lacewings and lady-bugs can help control aphid populations naturally.
2 **Pepper Spray:** A homemade pepper spray can deter aphids. Mix water with hot pepper flakes, strain, and spray on affected plants.
3 **Banana Peels:** Placing banana peels at the base of your plants can repel aphids. The peels release compounds that aphids find unpleasant.
4 **Essential Oils:** Essential oils like peppermint, rosemary, and clove can be used to repel aphids. Mix a few drops with water and spray on your plants.

Emotional and practical challenges of managing aphid infestations
Dealing with aphid infestations can be an emotional rollercoaster for gardeners. The frustration of seeing your hard work being undone by tiny pests can be overwhelming. However, with the right knowledge and tools, you can effectively manage aphid populations and keep your garden thriving.

Case study: The psychological impact of gardening challenges (Royal horticultural society)
The Royal Horticultural Society (RHS) has conducted studies on the psychological benefits of gardening and the emotional toll of pest infestations. Their research indicates that while gardening can significantly improve mental well-being, challenges such as aphid infestations can lead to stress and frustration. The RHS emphasizes the importance of community support, education, and resilience in overcoming these challenges.

> **Tips for Managing Gardening Frustration**
> - **Stay Positive:** Gardening is a learning experience. Don't let setbacks discourage you.
> - **Research:** Educate yourself on the pests and diseases that affect your plants.
> - **Community:** Join gardening groups or forums for support and advice.
> - **Patience:** Gardening takes time. Be patient and persistent.

Practical tips for managing aphid infestations

1 **Regular Monitoring:** Check your plants regularly for signs of aphids. Early detection is key to effective control.
2 **Watering Practices:** Proper watering can help keep plants healthy and less susceptible to aphid infestations.
3 **Plant Health:** Healthy plants are more resilient to pests. Ensure your plants receive adequate nutrients and care.
4 **Diverse Planting:** Planting a variety of species can help reduce the likelihood of aphid infestations.

Case study: Natural solutions in action

Case study: Natural aphid control in home gardens (Cornell university cooperative extension)

Cornell University Cooperative Extension provides resources and case studies on organic pest control methods for home gardens. One notable case study highlights the successful use of neem oil, insecticidal soap, and companion planting in a home garden in upstate New York.

Key Natural Pest Control Methods

- **Neem Oil:** A natural pesticide that disrupts aphid feeding and reproduction.
- **Insecticidal Soap:** Kills aphids on contact by dissolving their protective coating.
- **Companion Planting:** Using plants that repel aphids to protect vulnerable crops.
- **Beneficial Insects:** Introducing predators that naturally control aphid populations.

Natural vs. Organic Control Methods

- **Natural Methods:** These involve using substances and techniques derived from nature to control pests. Natural methods focus on non-synthetic, minimally processed ingredients like essential oils, plant extracts, and physical barriers. An example is using diatomaceous earth to deter insects or applying neem oil extracted from the neem tree.
- **Organic Methods:** Organic gardening methods adhere to standards set by organic certification bodies. These standards prohibit the use of synthetic fertilizers and pesticides, emphasizing ecological balance and sustainability. Organic methods include using compost for soil fertility, crop rotation to manage pests, and certified organic products like insecticidal soap or neem oil. While all organic methods are natural, not all natural methods meet organic certification criteria.

Conclusion

Aphids may be small, but their impact on home gardens can be significant. From ornamental plants like hibiscus to personal food gardens, these tiny pests pose a persistent challenge for gardeners. However, with the right knowledge and tools, you can manage aphid infestations effectively and organically.

This chapter has explored the personal battles gardeners face against aphids, sharing stories, practical advice, and tips for organic pest control.

By understanding the challenges and solutions, you can protect your garden and enjoy the fruits of your labor.

As we continue to learn and innovate, the future of gardening looks bright. With sustainable practices and a commitment to eco-friendly solutions, we can create healthy, vibrant gardens that are resilient to pests and diseases.

Glossary of Terms

Aphid-Resistant Plants: Crop varieties that possess natural defenses against aphid infestation.

Beneficial Insects: Insects that provide natural pest control by preying on or parasitizing harmful insect species.

Companion Planting: A method of planting different crops in proximity for pest control, pollination, and maximizing the use of space.

Insecticidal Soap: A soap-based pesticide that is effective against soft-bodied insects like aphids.

Neem Oil: A natural pesticide derived from the seeds of the neem tree, used to control a variety of pests.

Organic Gardening: A method of gardening that relies on natural substances and processes to grow plants, avoiding synthetic chemicals.

Parthenogenesis: A form of asexual reproduction where an organism produces offspring without fertilization.

Reflective Mulch: A type of mulch that reflects light, deterring pests like aphids from landing on plants.

Sticky Traps: Devices coated with a sticky substance used to trap flying insects.

Sustainable Agriculture: Farming practices that maintain and improve environmental health, economic profitability, and social equity.

Yellow Sticky Traps: Traps used to catch flying aphids and other pests, helping to monitor and reduce their populations.

Bibliography

Blackman, R. L., & Eastop, V. F. (2000). *Aphids on the World's Crops: An Identification and Information Guide.* Wiley.

Dixon, A. F. G. (1998). *Aphid Ecology: An Optimization Approach.* Chapman & Hall.

Makkouk, K. M., & Kumari, S. G. (2009). Viruses infecting pepper crops. *Advances in Virus Research, 75,* 1–35.

Raupp, M. J., & Shrewsbury, P. M. (2009). Sustainable pest management in urban landscapes. *Annual Review of Entomology, 54,* 233–250.

Simon, J. C., Rispe, C., & Sunnucks, P. (2002). Ecology and evolution of sex in aphids. *Trends in Ecology & Evolution, 17*(1), 34–39.

Van Emden, H. F., & Harrington, R. (2007). *Aphids as Crop Pests.* CABI.

CHAPTER 7

Human interaction and cultural significance

Introduction

Aphids might seem like an unlikely subject of cultural fascination, but these tiny sap-suckers have managed to leave their mark on human history and imagination. Often seen as mere pests in the garden, aphids have also been symbols, metaphors, and even muses for artists and writers. From their symbolic roles in folklore to their appearances in literature and art, aphids have intrigued and inspired people in surprising ways. They are not just garden nuisances but also actors in human storytelling, providing rich material for cultural expression.

In this chapter, we'll explore the cultural significance of aphids, examining how they have been represented and perceived throughout history. We'll journey through ancient myths, medieval superstitions, Renaissance art, and modern pop culture to uncover the diverse ways aphids have been woven into the human narrative. Whether they are symbols of destruction and resilience or characters in children's stories, aphids offer a unique lens through which to view our own culture and creativity.

Imagine a world where aphids are not just the pesky invaders of your prized roses but also the tiny heroes and villains of epic tales, the subjects of intricate paintings, and the metaphors in profound literary works. These small insects, often overlooked, have had a surprisingly significant role in shaping cultural expressions across different eras and societies.

In ancient times, aphids were woven into the fabric of mythology, representing everything from fertility and abundance to destruction and plague. They appeared in the stories that our ancestors told to explain the mysteries of nature and human existence. Fast forward to the Renaissance, and aphids

DOI: 10.1201/9781003537786-7

can be found hidden in the detailed botanical illustrations of artists striving to capture the natural world in all its complexity and beauty.

In literature, aphids have been used as symbols and metaphors, their roles ranging from the destructive forces in a garden to the resilient survivors in harsh conditions. Authors have drawn parallels between the behaviors of aphids and human traits, using these insects to add layers of meaning to their narratives.

Today, aphids continue to inspire, appearing in modern art and popular culture in ways that reflect our ongoing fascination with the natural world and its smallest inhabitants. They are featured in environmental art, highlighted in scientific documentaries, and even make cameo appearances in animated films and television shows.

By exploring these various facets of aphid representation, we gain insight into how human perceptions of these tiny creatures have evolved and what they reveal about our broader cultural and environmental attitudes. This chapter will take you on a journey through time and imagination, showing how aphids have been both despised and revered, trivialized and celebrated. Prepare to be surprised, amused, and perhaps even inspired by the cultural significance of aphids. So, buckle up and prepare to see aphids in a whole new light—as cultural icons.

Aphids in art

The depiction of insects in art has a long history, often symbolizing various human concerns, from the transient nature of life to the complexity of ecosystems. Aphids, though less celebrated than butterflies or bees, have found their way into artistic expressions.

Figure 7.1 Ants. (Shutterstock ID: 2182608729)

Renaissance intricacies

During the Renaissance, artists like Albrecht Dürer paid meticulous attention to detail in their works, capturing the natural world with incredible precision. While aphids themselves were not always the main subjects, the detailed flora in which they lived often featured these tiny insects. In Dürer's famous engraving "The Great Piece of Turf," the intricate rendering of plants subtly includes aphids, reflecting the artist's dedication to depicting nature accurately.

Modern interpretations

In contemporary art, aphids often symbolize the delicate balance of ecosystems and the unintended consequences of human actions. For example, the environmental artist BugLife created a series of installations highlighting the role of aphids in pollinator support systems, emphasizing their importance despite their status as pests.

Aphids in literature

Aphids might not be the stars of many literary works, but their presence in literature often serves to illustrate broader themes and add layers of meaning to narratives. From symbolic representations in political allegories to educational roles in children's books, aphids have found their way into the pages of various genres, each time enriching the story with their unique characteristics.

Symbolic aphids

Aphids are often used symbolically to represent parasitic and oppressive entities, illustrating how seemingly insignificant pests can have a significant impact on their surroundings. In George Orwell's "Animal Farm," aphids are used metaphorically to critique societal structures that allow certain groups to thrive at the expense of others. The mention of aphids in this political allegory serves to highlight the exploitation and parasitic nature of the ruling elite.

Orwell's use of aphids is subtle yet powerful. The aphids, like the corrupt leaders in the novel, feed off the hard work of others while contributing nothing of value themselves. This imagery reinforces the novel's themes of power, corruption, and the exploitation of the working class. The aphids' parasitic behavior mirrors the oppressive tactics used by the pigs to maintain control over the other animals on the farm.

In a more contemporary example, Margaret Atwood's dystopian novel "The Handmaid's Tale" uses aphids to symbolize the insidious nature of the theocratic regime. The brief mention of aphids infesting a garden in the Commander's household serves as a metaphor for the pervasive and hidden corruption within the seemingly idyllic society of Gilead. The aphids, like the regime, are difficult to eradicate and cause destruction from within, underscoring the novel's themes of oppression and resistance.

Children's literature

In children's literature, aphids often appear in stories about the natural world, teaching young readers about the complexities of ecosystems and the importance of each species, no matter how small. Eric Carle's "The Grouchy Ladybug" is a beloved example that introduces children to aphids as part of the food chain, helping them understand the interconnectedness of nature.

In "The Grouchy Ladybug," aphids are depicted as the primary food source for the titular character. Through colorful illustrations and engaging storytelling, Carle educates children about the role aphids play in the ecosystem. The ladybug's journey to find its next meal showcases the balance of predator and prey, emphasizing that even the smallest creatures have their place in the natural order.

Another example is Julia Donaldson's "What the Ladybird Heard," where aphids, though not central to the plot, make an appearance in the vibrant farmyard setting. The presence of aphids among the plants and animals subtly educates children about biodiversity and the intricate web of life in a farm ecosystem. The book's playful narrative and rhyming text make learning about nature fun and memorable for young readers.

Aphids as metaphors and plot devices

Aphids have also been used as metaphors and plot devices in various literary genres, adding depth and complexity to narratives. In detective fiction, for instance, an infestation of aphids can serve as a clue or a symbol of underlying decay. In Agatha Christie's "The Murder of Roger Ackroyd," the meticulous detective Hercule Poirot notices a garden infested with aphids, which he interprets as a sign of neglect and hidden turmoil within the household. This small detail becomes a metaphor for the larger secrets and deceit lurking beneath the surface.

In ecological literature, aphids often symbolize the delicate balance of nature and the consequences of human intervention. Rachel Carson's seminal work "Silent Spring" discusses the impact of pesticides on aphid populations and the resulting ecological imbalance. Carson's vivid descriptions of aphid infestations and their ripple effects on the environment serve as a powerful warning about the dangers of disrupting natural processes.

Modern literature and popular culture

In modern literature and popular culture, aphids continue to be used in creative and thought-provoking ways. In speculative fiction, aphids might be featured in futuristic scenarios where their behavior has been genetically modified for various purposes, ranging from ecological restoration to bio-warfare. Such imaginative portrayals highlight the potential for both beneficial and harmful applications of biotechnology.

In environmental novels, aphids are often depicted as indicators of ecological health. Authors use aphid populations to illustrate broader environmental themes, such as climate change and habitat destruction. For instance, in Barbara Kingsolver's "Flight Behavior," the protagonist observes changes in aphid activity as an early sign of shifting weather patterns and the larger impacts of global warming. Through these observations, Kingsolver connects the local experiences of her characters to global environmental issues.

Folklore Involving Aphids

Ancient Myths and Aphids In some ancient mythologies, aphids were believed to be the offspring of Aphrodite, the Greek goddess of love and beauty. This association with Aphrodite might seem odd, but it stems from the aphid's rapid reproduction and ability to appear suddenly, much like the unexpected nature of love.

Medieval Superstitions During the Middle Ages, aphids were often seen as omens. A sudden infestation could be interpreted as a sign of coming misfortune or a punishment for neglecting one's garden. Conversely, some cultures believed that aphids brought good luck, especially if they appeared on crops close to harvest time.

Aphids in Modern Culture

Pop culture references

In modern times, aphids have occasionally made their way into pop culture. In the animated film "A Bug's Life," aphids are depicted as livestock, humorously reversing the usual predator-prey dynamic by showing them being "milked" by ants.

Scientific popularity

Aphids also enjoy a certain level of fame in scientific communities and popular science media. Shows like BBC's "Life in the Undergrowth" feature aphids to highlight their complex interactions with ants and plants, showcasing the marvels of their tiny worlds.

Aphids in folklore and mythology

Aphids, despite their tiny size and often reviled presence in gardens and fields, have found their way into the folklore and mythology of various cultures. These narratives often imbue aphids with symbolic meanings, reflecting the complex relationship humans have with these insects. From symbols of fertility and abundance to harbingers of destruction and resilience, aphids have left a mark on cultural stories and traditions.

Symbolism in different cultures

Aphids have been interpreted in various ways across different cultures, often symbolizing themes of fertility, destruction, and resilience. These interpretations provide insight into how different societies viewed the natural world and their place within it.

Fertility and abundance

In some agricultural societies, aphids were seen as symbols of fertility and abundance due to their rapid reproduction rates. Farmers, recognizing the prolific nature of these insects, sometimes performed rituals to appease them, hoping to harness their prolific nature to ensure bountiful harvests.

Ancient Egyptian mythology

In Ancient Egypt, the land's fertility was paramount, and any creature associated with prolific reproduction was seen in a somewhat positive light. While aphids were pests, their ability to multiply rapidly was viewed as a symbol of the land's potential to produce abundant crops. Some farmers believed that leaving a portion of their fields to aphids would appease these insects, thus ensuring the rest of their crops would flourish. This practice, though not scientifically sound, reflected a deep-seated desire to maintain harmony with all elements of nature, even the pests.

Celtic folklore

In Celtic folklore, aphids were sometimes linked to the goddess Brigid, a deity associated with fertility, agriculture, and healing. Aphids, due to their rapid and prolific breeding, were seen as tiny emissaries of growth and renewal. During the festival of Imbolc, which marks the beginning of spring, it was believed that seeing aphids on plants was a sign of the coming abundance and the earth's renewed fertility. While not celebrated, their presence was accepted as a natural part of the cycle of life.

Chinese symbolism

In traditional Chinese symbolism, insects often carry significant meanings. Aphids, with their ability to produce numerous offspring quickly, were sometimes seen as symbols of abundance and prosperity. The Chinese word for aphid, "虱" (shī), while generally referring to lice, can also metaphorically represent small creatures that multiply quickly. In some rural areas, farmers would leave small offerings in their fields, a practice meant to respect and appease the tiny creatures that played a role in the agricultural ecosystem.

Destruction and resilience

Conversely, aphids were also seen as harbingers of destruction. Their sudden appearance and ability to devastate crops made them symbols of nature's uncontrollable forces. However, this destructive capacity also underscored a message of resilience—both for the aphids and the humans who had to constantly adapt to their presence.

Medieval European folklore

In medieval Europe, aphids were often viewed with dread due to their potential to destroy entire crops. They were seen as omens of famine and hardship, with infestations interpreted as divine punishment or a test of endurance. Stories from this era often depict farmers battling against these relentless invaders, using prayers and charms to ward them off. The aphids' tenacity and the farmers' continual efforts to combat them underscored a larger narrative of human resilience in the face of natural adversities.

Humorous take on aphids in culture

Who knew that aphids, those tiny, often reviled garden pests, could also be the source of laughter and humor? Despite their small size and the general annoyance they cause to gardeners worldwide, aphids have managed to wiggle their way into the lighter side of popular culture. Let's dive into the comedic side of aphids and see how these insects have become the butt of many jokes and humorous stories.

The comedic side of aphids

Aphids might be more famous for their destructive habits and rapid population growth, but they've also found a niche in humor, becoming a quirky subject in various forms of media. Their rapid reproduction, unique behaviors, and sheer numbers make them perfect material for jokes and funny anecdotes.

Aphids in pop culture

"The big bang theory"

In the popular television show "The Big Bang Theory," one of the characters, an entomologist named Dr. Beverly Hofstadter, makes a joke about aphids. She explains how their rapid reproduction rates make them the "rabbits of the insect world." This analogy, while humorous, also underscores the aphid's notoriety for rapid population growth. The joke lands well with the show's audience, many of whom can relate to the idea of pests multiplying out of control.

Monty python's flying circus

In an episode of Monty Python's Flying Circus, a sketch humorously portrays a society where aphids have taken over, much to the dismay of the human characters. The exaggerated reactions and over-the-top scenarios highlight the absurdity of aphid infestations while poking fun at humanity's often futile attempts to control nature. The sketch serves as a reminder of how even the smallest creatures can become a significant nuisance.

Saturday night live

Saturday night live once featured a parody commercial for a fictional product called "Aphid-B-Gone," a ridiculous contraption that promises to eradicate aphids from gardens in the most absurd ways. The commercial's over-the-top claims and the actors' exaggerated enthusiasm create a hilarious take on the lengths people will go to combat these tiny pests. It's a perfect example of how humor can be used to address everyday problems in a lighthearted manner.

Fun facts and trivia

Even beyond television and film, aphids have inspired a host of fun facts and trivia that often verge on the absurd. These little tidbits not only entertain but also educate, highlighting some of the more bizarre aspects of aphid biology and behavior.

Aphid facts

- **Aphid Teleportation (Sort of):** Aphids can seemingly appear out of nowhere due to their parthenogenetic reproduction, where females give birth to live young without mating. This can lead to sudden, massive population explosions, leaving gardeners scratching their heads and wondering where all these aphids came from.
- **Aphid-Alien Connection:** Some conspiracy theorists jokingly claim that aphids are evidence of extraterrestrial life because of their alien-like reproductive strategies and rapid population growth. While this is all in good fun, it does make one marvel at just how strange and efficient aphid reproduction is.
- **Tiny Sap Vampires:** Aphids are often referred to as "sap vampires" because they feed on the sap of plants, much like a vampire feeds on blood.

This humorous nickname underscores the aphid's parasitic nature and their ability to drain the life out of plants.

Funny Garden Fails

Gardening is a hobby that often comes with its fair share of humorous mishaps, especially when aphids are involved. Here are a few laugh-out-loud garden fails that highlight the unexpected challenges aphids can bring to the gardening experience.

The aphid avalanche

One gardener recalls the time they sprayed their rosebushes with a high-powered hose to dislodge aphids. The plan worked—perhaps too well. The aphids came off in such numbers that the gardener described it as an "aphid avalanche." The sight of countless aphids pouring off the plants and the subsequent scramble to avoid the insect deluge is a story that gets retold at every garden party.

The ladybug invasion

Another gardener decided to introduce ladybugs to their garden to combat an aphid infestation. Excitedly, they released thousands of ladybugs, expecting them to immediately start munching on aphids. Instead, the ladybugs dispersed in every direction, some even ending up inside the gardener's house. The image of a home suddenly filled with thousands of ladybugs, while amusing in hindsight, was a humorous yet slightly overwhelming experience.

The garlic spray fiasco

An enthusiastic gardener read that garlic spray could deter aphids. Armed with a homemade concoction, they liberally sprayed their entire garden. Unfortunately, they didn't strain the mixture properly, resulting in clogged sprayers and garlic chunks covering the plants. The aphids seemed undeterred, and the garden smelled like an Italian restaurant for weeks.

Conclusion

Aphids may be tiny, but their impact on human culture is surprisingly significant. From their roles in art and literature to their symbolic meanings in folklore, aphids have intrigued and inspired people throughout history. By exploring the cultural significance of aphids, we gain a deeper appreciation for these tiny insects and their place in the human narrative.

Throughout the ages, aphids have been more than just garden pests. They have been symbols of fertility, resilience, and destruction, appearing in the myths and stories of various cultures. Their prolific reproduction has made them emblems of abundance and natural vitality, while their capacity to devastate crops has underscored the theme of nature's uncontrollable power.

In literature, aphids have served as metaphors and symbols, adding layers of meaning to narratives. Whether they are representing oppressive societal structures or teaching children about the food chain, aphids have proven to be versatile and impactful characters in written works. Their presence in stories from George Orwell's "Animal Farm" to Eric Carle's "The Grouchy Ladybug" demonstrates their ability to convey complex themes and lessons.

Art, too, has found a place for aphids. From the intricate botanical illustrations that detail their forms and functions to the more whimsical portrayals in modern media, aphids have been depicted in ways that highlight both their beauty and their nuisance. These artistic representations help us see aphids not just as pests but as integral parts of the natural world with their own roles and stories.

The humorous takes on aphids, such as those seen in pop culture and comedic sketches, provide much-needed comic relief and allow us to laugh at the absurdity of battling these tiny foes. Shows like "The Big Bang Theory" and sketches from "Monty Python's Flying Circus" remind us that even the smallest creatures can have a big impact on our lives, and sometimes the best way to deal with adversity is to find humor in it.

By examining the cultural significance of aphids, we not only understand their ecological roles but also recognize their place in the tapestry of human experiences. Aphids have been woven into the fabric of our stories, our art, and our humor, making them more than just insects—they are characters in the grand narrative of life.

As we reflect on the many ways aphids have influenced human culture, we can appreciate the intricate connections between all living things. This exploration of aphids in culture enriches our understanding of the natural world and our place within it. It reminds us that even the smallest creatures can inspire big ideas and that the stories we tell about them reveal much about ourselves.

In the end, aphids, with their resilience and adaptability, mirror the human spirit. Their ability to thrive in the face of adversity, to find new ways to survive and multiply, and to leave a lasting mark on the world around them is a testament to the interconnectedness of all life. So next time you see an aphid, consider not just its impact on your garden but its place in the broader narrative of human culture and nature.

Glossary of Terms

Aphid: Small sap-sucking insects that are common pests in gardens and agriculture.

Biological Control: The use of natural predators or parasites to manage pest populations.

Companion Planting: A method of planting different crops in proximity for pest control, pollination, and maximizing use of space.

Honeydew: A sugary liquid secreted by aphids and other plant-sucking insects, often harvested by ants.

Insecticide: A type of pesticide specifically used to kill or control insects.

Mutualism: A symbiotic relationship where both parties benefit.

Organic Gardening: A method of gardening that relies on natural substances and processes to grow plants, avoiding synthetic chemicals.

Parthenogenesis: A form of asexual reproduction where an organism produces offspring without fertilization.

Pesticide: A chemical substance used to kill pests, including insects and weeds.

Sticky Traps: Devices coated with a sticky substance used to trap flying insects.

Bibliography

Blackman, R. L., & Eastop, V. F. (2000). *Aphids on the World's Crops: An Identification and Information Guide.* Wiley.

Carle, E. (1977). *The Grouchy Ladybug.* HarperCollins.

Dixon, A. F. G. (1998). *Aphid Ecology: An Optimization Approach.* Chapman & Hall.

Raupp, M. J., & Shrewsbury, P. M. (2009). Sustainable pest management in urban landscapes. *Annual Review of Entomology, 54,* 233–250.

Simon, J. C., Rispe, C., & Sunnucks, P. (2002). Ecology and evolution of sex in aphids. *Trends in Ecology & Evolution, 17*(1), 34–39.

Van Emden, H. F., & Harrington, R. (2007). *Aphids as Crop Pests.* CABI.

CHAPTER 8

Economic impact of aphids

Introduction

Imagine this: you're a farmer standing in the middle of your lush, green fields, marveling at the promise of a bountiful harvest. The sun is shining, the soil is rich, and your crops are flourishing. You've invested countless hours and resources into ensuring this season's success. Suddenly, you notice a fine, sticky substance coating the leaves and stems of your plants. Your heart sinks as you see a swarm of tiny insects—aphids—feasting on your crops. Welcome to the world of aphids, where these seemingly insignificant bugs can wreak havoc on your agricultural investments.

Aphids may be small, but their impact on agriculture is monumental. These tiny, sap-sucking pests can cause direct damage by feeding on plant tissues, which weakens plants, stunts their growth, and reduces yields. But the destruction doesn't stop there. Aphids are also notorious vectors for plant viruses, transmitting diseases that can further devastate crops. The economic losses resulting from aphid infestations are substantial, affecting not only individual farmers but entire agricultural sectors.

In this chapter, we explore the financial repercussions of aphid infestations, the costs associated with managing these pests, and the broader economic strategies designed to keep them in check. We'll delve into the direct and indirect impacts of aphids on crop production, examining case studies from around the world to illustrate the scale of the problem. We'll also look at the various pest management techniques employed by farmers, from traditional chemical insecticides to innovative biological controls and integrated pest management (IPM) strategies. By understanding the extent

DOI: 10.1201/9781003537786-8

of the problem and the measures in place to combat it, we can appreciate the significant economic burden these tiny insects impose on farmers and agricultural systems worldwide.

Picture this scenario: A farmer in Iowa, anticipating a record-breaking soybean harvest, suddenly faces an aphid invasion. The plants, once lush and vibrant, are now covered with the sticky residue of honeydew, a tell-tale sign of aphid activity. Leaves curl and turn yellow, pod formation is disrupted, and yields plummet. This scenario isn't dystopian fiction—it's a common reality for many farmers. The financial implications are immediate and severe, encompassing not just the loss of crops but the cost of emergency pest control measures and the long-term impact on soil health and future harvests.

Aphid infestations require farmers to pivot swiftly, often incurring significant expenses to purchase and apply insecticides. These costs are compounded by the labor required for application and the potential environmental consequences of chemical use. Moreover, aphids have a notorious ability to develop resistance to insecticides, necessitating the use of more potent—and often more expensive—chemicals.

However, the financial toll isn't limited to the direct costs of pest control. Aphids' role as vectors for plant viruses can lead to widespread disease outbreaks, further reducing yields and quality of produce. In some cases, entire fields may be written off, with the economic ripple effect extending beyond the farm to affect local economies and food supply chains.

This chapter will also explore the broader economic strategies employed by policymakers and agricultural organizations to support farmers in their battle against aphids. Government subsidies, research funding for developing resistant crop varieties, and extension services providing education and resources are all part of a multifaceted approach to mitigating the economic impact of aphids. International cooperation and knowledge sharing are crucial as aphid infestations are a global issue requiring a concerted effort.

To truly grasp the economic impact of aphids, we'll examine various case studies and real-world examples. From the pea aphid epidemic in the United Kingdom to aphid-resistant crop initiatives in the United States, these stories highlight the ongoing struggle between farmers and these persistent pests. We'll also look at the future of pest management, exploring cutting-edge technologies and sustainable practices that hold promise for reducing the economic burden of aphids.

By the end of this chapter, you'll have a comprehensive understanding of how aphids affect the agricultural economy, the strategies in place to combat them, and the ongoing research aimed at finding more effective and sustainable solutions. Despite their size, aphids wield considerable power in

Figure 8.1 Winged aphids. (Shutterstock ID: 2177850555)

Figure 8.2 Roses. (Shutterstock ID: 2467661865)

Figure 8.3 Lacewing. (Shutterstock ID: 1067407361)

the agricultural world, and understanding their impact is crucial for ensuring the resilience and sustainability of our food systems.

The economic toll of aphid damage

Aphids feed by piercing plant tissues and sucking out sap, which not only weakens the plant but also stunts its growth. The direct damage they cause can lead to reduced yields and lower-quality produce, both of which have significant economic implications for farmers.

Case study: The pea aphid epidemic

In the late 1990s, the United Kingdom experienced a severe outbreak of pea aphids (*Acyrthosiphon pisum*), which devastated pea crops across the country. This infestation led to substantial financial losses for farmers and the agricultural industry as a whole. The pea aphid infestation caused direct damage to the plants by sucking sap and stunting growth, resulting in reduced pod formation and lower yields. Additionally, the aphids transmitted several plant viruses, exacerbating the damage and further reducing crop quality and quantity.

The economic impact was staggering. Farmers faced increased costs due to the need for more frequent and intensive pest control measures, including chemical insecticides and biological controls. The combined effect of reduced yields and increased pest management costs led to financial losses estimated at over £200 million for the UK pea industry during the peak years of the infestation.

Key Statistics on Aphid Damage
- **Global Crop Losses:** Aphids are responsible for approximately 10–15% of global crop losses annually.
- **Economic Impact in the United Kingdom:** The pea aphid epidemic caused over £200 million in annual losses.
- **Yield Reductions:** Aphid infestations can reduce crop yields by 20–50%, depending on the severity.

The cost of pest management

Managing aphid populations is a costly endeavor. Farmers must invest in various pest control methods to protect their crops, from chemical insecticides to biological controls and IPM strategies.

Chemical insecticides

Chemical insecticides are often the first line of defense against aphids. While effective, they come with significant costs, both financial and environmental. The purchase and application of insecticides can be expensive, and repeated use can lead to the development of resistant aphid populations, necessitating even more expensive and potent chemicals.

Biological controls

Biological controls, such as introducing natural predators like ladybugs and lacewings, offer a more sustainable approach to managing aphid populations. However, the initial investment in these biological agents and the ongoing management required to maintain their populations can still be costly.

Integrated pest management

IPM combines chemical, biological, and cultural practices to manage pest populations in an economically and environmentally sustainable way. While IPM can reduce long-term costs and environmental impact, the initial setup and ongoing monitoring can be resource-intensive.

Economic Analysis of Pest Management Costs
- **Insecticide Costs:** Farmers can spend anywhere from $20 to $100 per acre on chemical insecticides.
- **Biological Control Costs:** Introducing natural predators can cost between $50 and $200 per acre, depending on the species and management practices.
- **IPM Costs:** Implementing IPM strategies can range from $50 to $150 per acre, but can reduce long-term expenses.

Economic strategies for mitigation
To mitigate the economic impact of aphids, farmers and policymakers employ various strategies, from government subsidies to research and development in pest-resistant crop varieties.

Government subsidies and support
Governments often provide financial support to farmers affected by aphid infestations. Subsidies can help cover the costs of pest management and compensate for lost yields. Additionally, government-funded research programs focus on developing new pest control methods and resistant crop varieties.

Research and development
Investing in research and development is crucial for finding long-term solutions to aphid infestations. Scientists are working on developing genetically modified crops that are resistant to aphids and improving biological control methods to make them more cost-effective and efficient.

Farmer education and training
Educating farmers about the best practices for managing aphid populations can significantly reduce the economic impact of these pests. Extension services and agricultural organizations offer training programs on IPM, the use of biological controls, and sustainable farming practices.

Case study: Aphid-resistant crop varieties
Researchers at agricultural universities have made significant strides in developing aphid-resistant crop varieties. For example, the development of aphid-resistant wheat and barley varieties has provided farmers with new tools to combat these pests. These resistant varieties not only reduce the need for chemical insecticides but also help ensure stable yields, protecting farmers' incomes.

The broader economic impact
The economic impact of aphids extends beyond individual farms. Widespread infestations can affect entire agricultural sectors, leading to higher food prices, reduced export revenues, and increased costs for consumers.

Impact on food prices
Aphid infestations can lead to reduced crop yields and lower-quality produce, which in turn drives up food prices. Consumers may pay more for fresh produce, processed foods, and other agricultural products affected by aphid damage.

Reduced export revenues

Countries that rely heavily on agricultural exports can see a significant impact on their economies due to aphid infestations. Reduced yields and lower-quality crops can lead to decreased export revenues, affecting the overall economic health of the nation.

Increased costs for consumers

The increased costs of managing aphid infestations are often passed on to consumers. Higher prices for food and agricultural products can strain household budgets and reduce disposable income, affecting overall economic stability.

Key Economic Impacts of Aphid Infestations

- **Food Prices:** Increased costs for fresh produce and processed foods.
- **Export Revenues:** Decreased revenues from agricultural exports.
- **Consumer Costs:** Higher prices for food and agricultural products.

Future directions in economic mitigation

As the global population continues to grow, the demand for food and agricultural products will increase, making effective aphid management even more critical. Future economic strategies for mitigating the impact of aphids will likely focus on sustainable practices, technological advancements, and international cooperation.

Sustainable agricultural practices

Promoting sustainable agricultural practices can help reduce the economic impact of aphids. Practices such as crop rotation, intercropping, and the use of cover crops can enhance biodiversity and reduce the likelihood of severe aphid infestations.

Technological advancements

Advancements in technology, such as precision agriculture and drone monitoring, can help farmers detect and manage aphid populations more effectively. These technologies can reduce the need for chemical insecticides and lower the overall costs of pest management.

International cooperation

Aphid infestations are a global issue, and international cooperation is essential for finding long-term solutions. Sharing research, resources, and best practices can help countries develop effective strategies for managing aphid populations and mitigating their economic impact.

Case study: International collaboration in aphid management

An international consortium of researchers and agricultural experts has been working together to develop IPM strategies for aphids. By sharing data, resources, and expertise, this collaboration has led to significant advancements in biological control methods, pest-resistant crop varieties, and sustainable farming practices.

Conclusion

Aphids, despite their tiny size, have a substantial economic impact on agriculture. From direct crop damage to the costs of pest management, these pests pose significant financial challenges for farmers and policymakers. By understanding the economic toll of aphid infestations and employing effective management strategies, we can mitigate their impact and ensure the stability of our agricultural systems.

Through case studies, economic analysis, and practical insights, this chapter has explored the true cost of aphids on agriculture. By investing in research, promoting sustainable practices, and fostering international cooperation, we can develop long-term solutions to protect our crops and our economies from the persistent threat of aphids.

Glossary of Terms

Aphid-Resistant Plants: Crop varieties that possess natural defenses against aphid infestation.

Biological Control: The use of natural predators or parasites to manage pest populations.

Chemical Insecticides: Synthetic chemicals used to kill or repel insects.

Integrated Pest Management (IPM): A sustainable approach to managing pests that combines biological, chemical, and cultural methods.

Parthenogenesis: A form of asexual reproduction where an organism produces offspring without fertilization.

Precision Agriculture: Farming practices that use technology to monitor and manage crops with high precision.

Sustainable Agriculture: Farming practices that maintain and improve environmental health, economic profitability, and social equity.

Bibliography

Blackman, R. L., & Eastop, V. F. (2000). *Aphids on the World's Crops: An Identification and Information Guide*. Wiley.

Dixon, A. F. G. (1998). *Aphid Ecology: An Optimization Approach*. Chapman & Hall.

Makkouk, K. M., & Kumari, S. G. (2009). Viruses infecting pepper crops. *Advances in Virus Research*, 75, 1–35.

Simon, J. C., Rispe, C., & Sunnucks, P. (2002). Ecology and evolution of sex in aphids. *Trends in Ecology & Evolution*, 17(1), 34–39.

Van Emden, H. F., & Harrington, R. (2007). *Aphids as Crop Pests*. CABI.

CHAPTER 9

Tomorrow land: Future directions

Introduction

Picture this: It's the year 2040. You're strolling through your advanced, tech-driven farm, surrounded by a symphony of robotic drones, AI-driven sensors, and bioengineered crops. Your fields are a testament to modern agricultural science, with plants engineered to resist drought, pests, and disease. Life seems idyllic as you marvel at the seamless blend of technology and nature until you spot them—those pesky aphids, still thriving, still causing mayhem. Will these tiny insects ever be eradicated, or are we doomed to coexist with them forever?

In the future, when the battle against aphids continues, we are equipped with new tools and strategies at our disposal. Despite the leaps and bounds we've made in agricultural technology, aphids remain a persistent threat. These minute invaders have an uncanny ability to adapt, survive, and reproduce, making them formidable adversaries. But fear not! The arsenal of modern agriculture is more sophisticated than ever, offering hope for more effective and sustainable pest management.

In this chapter, we'll dive into the potential impacts of environmental changes on aphid populations and explore the cutting-edge technologies and research that promise to revolutionize pest management. From the rise of artificial intelligence and genetic engineering to the role of international cooperation and sustainable farming practices, we'll speculate on the future directions of aphid control. Buckle up and join us on this speculative journey, where science fiction meets the gritty reality of agricultural pest control. It's a world where the fight against aphids is waged not just with pesticides but with precision agriculture, biological controls, and innovative strategies that could change the face of farming forever.

DOI: 10.1201/9781003537786-9

Imagine drones flitting across your fields, their sensors detecting the slightest hint of aphid activity. These drones are equipped with AI algorithms that can analyze the data in real-time, identifying the best course of action. They might release a targeted spray of biopesticides, drop a payload of ladybugs, or even deploy tiny robots designed to remove aphids by hand (or rather, by claw). It's a scene straight out of a sci-fi novel, but it's closer to reality than you might think.

As we explore these advancements, we'll also consider the broader implications of environmental changes. Climate change, with its shifting temperatures and precipitation patterns, could alter aphid behaviors and population dynamics in unpredictable ways. Will warmer winters lead to larger aphid populations in the spring? Will increased rainfall help or hinder their spread? These are questions that researchers are only beginning to answer.

Furthermore, we'll delve into the role of genetic engineering in creating crops that are naturally resistant to aphids. By understanding the genetic mechanisms that make some plants more vulnerable than others, scientists can develop new varieties that deter aphids naturally, reducing the need for chemical interventions.

But technological advancements aren't the only focus. We'll also examine the economic strategies and policies that support sustainable pest management practices. Governments and international organizations play a crucial role in funding research, providing subsidies, and educating farmers about the best practices for managing aphid populations.

So, whether you're a farmer looking for practical solutions, a researcher interested in the latest scientific developments, or simply a curious reader fascinated by the intersection of technology and agriculture, this chapter offers a comprehensive look at the future of aphid control. Together, we'll explore how humanity can outsmart one of nature's most resilient pests and ensure a more sustainable and productive future for agriculture.

Get ready to journey into the future of farming, where the age-old battle against aphids meets the cutting-edge innovations of tomorrow. Let's envision a world where our fields are not just battlegrounds but laboratories of progress, where every challenge is met with ingenuity and every pest is just another problem to solve. Welcome to the future of aphid management—it's going to be an exciting ride.

Emerging technologies in pest control

The age of the drones

In the future, our skies will be filled with more than just birds and planes. Picture fleets of drones patrolling fields, equipped with advanced imaging systems and AI algorithms capable of detecting aphid infestations in real-time. These drones can deliver precise doses of biopesticides or release natural predators like ladybugs directly onto affected plants. It's like having a team of mini air marshals dedicated to keeping your crops aphid-free.

The idea of drones in agriculture isn't entirely new, but their potential has only just begun to be realized. With advancements in AI and machine learning, these flying gadgets are becoming smarter and more efficient. They can survey large areas quickly, process data on the fly, and respond to pest threats with pinpoint accuracy. Imagine a scenario where instead of blanketing an entire field with chemicals, drones selectively target only the infested areas, reducing chemical use and environmental impact.

Case study: The drone wars

In a groundbreaking pilot project, farmers in California's Central Valley employed drones to manage aphid populations in their tomato fields. The drones, equipped with high-resolution cameras and machine learning software, identified hotspots of aphid activity. They then deployed biocontrol agents, significantly reducing the aphid population and minimizing the need for chemical pesticides. The economic benefits included lower pest management costs and increased crop yields.

The project, a collaboration between agricultural scientists, drone technology companies, and local farmers, aimed to test the efficacy of drone-based pest management. The drones were equipped with a variety of sensors, including multispectral cameras to detect changes in plant health indicative of aphid infestation.

As the drones flew over the fields, their cameras captured high-resolution images that were fed into an AI system. This system had been trained to recognize the signs of aphid damage, such as yellowing leaves and the presence of honeydew. The AI could differentiate between aphid damage and other plant health issues, ensuring accurate identification.

Upon detecting an aphid hotspot, the drones deployed their payloads of biocontrol agents—natural predators like ladybugs and lacewings. These insects were released directly onto the affected plants, where they immediately began preying on the aphids. In cases where biocontrol was insufficient, the drones applied precise doses of biopesticides, targeting only the infested areas.

The results were impressive. Farmers saw a significant reduction in aphid populations without the widespread use of chemical pesticides. This not only lowered pest management costs but also increased crop yields by maintaining healthier plants. Moreover, the reduced reliance on chemicals meant less environmental impact and lower a risk of pesticide resistance developing in aphid populations.

The role of AI and machine learning

The integration of AI and machine learning is crucial to the success of drone-based pest management. These technologies enable drones to not just see but understand what they're looking at. Machine learning algorithms can process vast amounts of data, learning to identify subtle signs of aphid infestations that might be missed by the human eye.

Beyond real-time detection, AI can also predict future pest outbreaks. By analyzing historical data, weather patterns, and current field conditions, AI systems can forecast when and where aphid infestations are likely to occur. This allows farmers to take preventive measures, such as adjusting planting schedules or preemptively deploying biocontrol agents.

The beauty of machine learning is that it gets better over time. As drones gather more data and as algorithms are refined, the accuracy and efficiency of these systems will continue to improve. This creates a feedback loop where each season of data collection and pest management makes the system more effective.

Practical considerations and challenges

While the potential benefits of drone technology are immense, there are practical challenges to consider. Drones require significant initial investment, and their operation demands technical expertise. There are also regulatory hurdles related to airspace use and privacy concerns.

Cost vs. benefit

The cost of deploying drone technology can be a barrier for small-scale farmers. However, as the technology becomes more widespread and competition increases, prices are expected to drop. Moreover, the long-term savings from reduced chemical use and increased crop yields can offset the initial investment.

Training and maintenance

Operating drones and interpreting their data requires specialized skills. Farmers and agricultural workers need training to effectively use these systems. Additionally, regular maintenance of drones is essential to ensure their reliability and performance.

Regulatory framework

The use of drones in agriculture is subject to regulations that vary by country and region. Farmers must navigate these regulations, which can include restrictions on flight altitudes, times, and areas. Working with local authorities and staying informed about legal requirements is crucial for compliance.

The age of the drones represents a significant leap forward in pest management. As technology advances, we can expect drones to become even more integral to agriculture, offering precise, efficient, and sustainable solutions to the age-old problem of aphid infestations.

Looking ahead, we might see the development of autonomous drone swarms that can communicate and collaborate to manage entire farms. Advances in battery technology could extend flight times, while improvements in sensor technology could provide even more detailed data.

Drones are just one part of a broader trend toward integrated agricultural systems. Combining drone technology with ground-based sensors, automated irrigation systems, and smart machinery can create a cohesive, efficient farm management system. This holistic approach can optimize every aspect of farming, from planting and watering to pest control and harvesting.

Global implications

As drone technology becomes more accessible, its benefits could extend to developing countries where traditional pest management methods are less effective. By providing affordable, efficient solutions, drones can help improve food security and agricultural productivity worldwide.

The future of aphid management is taking to the skies. Drones, powered by AI and machine learning, offer a promising solution to one of agriculture's most persistent problems. While challenges remain, the potential benefits for farmers, the environment, and global food security are immense. As we continue to innovate and refine these technologies, the vision of a future where aphids are kept in check by fleets of flying guardians becomes increasingly attainable.

Genetic engineering and biocontrol

As genetic engineering technologies advance, we may soon see crops that are entirely resistant to aphids. By tweaking the DNA of plants to produce natural aphid repellents or toxins, scientists aim to create a future where aphid infestations are a thing of the past.

In the world of pest management, genetic engineering and biocontrol are emerging as powerful allies. Imagine a future where your crops are not just surviving aphid attacks but are entirely immune to these tiny, sap-sucking invaders. This is not a far-fetched dream but a tangible goal that scientists are working towards with remarkable zeal. By tweaking the DNA of plants

> **Genetic Engineering Marvels**
>
> Imagine tomatoes that can scream at aphids (okay, maybe not scream, but release a deterrent chemical at the first sign of an aphid invasion). This genetic modification could reduce the reliance on chemical pesticides and pave the way for more sustainable farming practices.

to produce natural aphid repellents or toxins, researchers aim to create a future where aphid infestations are a thing of the past.

The science of genetic engineering

Genetic engineering involves modifying the genetic material of an organism to introduce desirable traits. In the context of agriculture, this means creating plants that can fend off pests without the need for external interventions like chemical pesticides.

Achieving Pest Resistant Plants

1   **Incorporating Natural Repellents:** Some plants naturally produce compounds that repel aphids. By identifying and isolating the genes responsible for these compounds, scientists can introduce them into crops that are typically vulnerable to aphid attacks.
2   **Producing Aphid Toxins:** Certain plants and microorganisms produce toxins that are lethal to aphids but harmless to humans and other animals. By engineering crops to produce these toxins, researchers can create self-defending plants that kill aphids on contact.
3   **Enhancing Plant Defenses:** Plants have their own immune systems, albeit simpler than those of animals. Genetic engineering can enhance these natural defenses, making plants more resilient to aphid attacks. This can involve bolstering the production of defensive chemicals or strengthening physical barriers like thicker cell walls.

Case study: Genetically engineered wheat

One of the promising areas of research is the development of genetically engineered wheat that is resistant to aphids. In a recent study, scientists introduced genes from wild relatives of wheat into commercial varieties. These genes enabled the wheat to produce higher levels of a natural repellent called benzoxazinoid, which deters aphids from feeding. Fields planted with genetically engineered wheat showed a significant reduction in aphid populations compared to non-engineered fields. Farmers reported using fewer chemical pesticides, which translated to cost savings and reduced environmental impact. Healthier, aphid-free plants resulted in higher yields, improving the overall productivity of the farms.

Biocontrol: Nature's pest management

While genetic engineering is the high-tech approach to pest management, biocontrol represents the natural, eco-friendly strategy. Biocontrol involves using natural predators, parasites, or pathogens to control pest populations. It's like enlisting Mother Nature's army to fight off the invaders.

Biocontrol Basics

1 **Natural Predators:** Ladybugs, lacewings, and parasitic wasps are all-natural enemies of aphids. These beneficial insects can be introduced into fields to keep aphid populations in check.

2 **Pathogens:** Certain fungi, bacteria, and viruses specifically target aphids. These biological agents can be used to infect and kill aphid colonies without harming the crops or other beneficial insects.

3 **Habitat Management:** Creating an environment that supports the natural enemies of aphids can help keep these pests under control. This might involve planting cover crops that attract beneficial insects or maintaining hedgerows and other habitats around fields.

Figure 9.1 Ladybug lacewing. (Shutterstock ID: 413346148)

Case study: The use of ladybugs in California vineyards

In the vineyards of California, grape growers have turned to ladybugs as a natural solution to aphid infestations. Ladybugs are voracious aphid eaters, capable of consuming hundreds of aphids in their lifetime. By releasing thousands of ladybugs into their vineyards, growers have successfully managed aphid populations without resorting to chemical pesticides. The introduction of ladybugs led to a significant decrease in aphid numbers, protecting the grapevines from damage. Reduced pesticide use lowered costs for the growers and minimized the environmental impact on the surrounding ecosystems. The presence of ladybugs and other beneficial insects contributed to greater biodiversity within the vineyards, promoting a healthier and more resilient agricultural system.

The future of genetic engineering and biocontrol

The potential of genetic engineering and biocontrol in agriculture is immense. As these technologies continue to advance, we can expect to see even more innovative solutions for managing aphid populations and other pests. The integration of these approaches offers a sustainable, long-term strategy for protecting crops and ensuring food security.

Challenges and considerations

While the promise of genetic engineering and biocontrol is exciting, there are challenges and considerations to keep in mind:

1 **Public Perception:** Genetic modification of crops can be a contentious issue, with concerns about safety and environmental impact. Public education and transparent research are crucial to gaining acceptance.
2 **Regulatory Hurdles:** The development and deployment of genetically engineered crops and biocontrol agents are subject to strict regulations. Navigating these regulations can be time-consuming and costly.
3 **Ecological Balance:** Introducing new species or altering existing ones can have unforeseen consequences on the ecosystem. Careful monitoring and adaptive management are essential to avoid disrupting ecological balance.

As genetic engineering technologies advance, we may soon see crops that are entirely resistant to aphids. By tweaking the DNA of plants to produce natural aphid repellents or toxins, scientists aim to create a future where aphid infestations are a thing of the past. Coupled with biocontrol methods that leverage nature's own pest management strategies, we have a powerful arsenal to combat aphid populations sustainably and effectively.

The future of pest management is bright, with the potential for reduced chemical use, lower costs, and healthier crops. Through continued research, innovation, and collaboration, we can achieve a balanced and sustainable approach to agriculture, ensuring food security for generations to come.

Environmental changes and aphid populations

Climate change: The aphid utopia?

As global temperatures rise and weather patterns shift, aphid populations are likely to respond in kind, potentially turning our current challenges with these tiny pests into a full-blown agricultural crisis. Welcome to what might become the Aphid Utopia—a world where climate change creates the perfect conditions for these sap-sucking insects to thrive.

Predictions for Aphid Populations

- **Warmer Winters:** Expect larger aphid populations emerging earlier in the spring.
- **Increased Rainfall:** Could either drown aphids or promote lush plant growth, providing more food sources.
- **Extended Growing Seasons:** More opportunities for aphids to reproduce and spread.

Warmer winters: The ultimate aphid winter getaway

One of the most significant impacts of climate change on aphid populations is the effect of warmer winters. Typically, the cold season serves as a natural population control mechanism, killing off a significant portion of overwintering eggs and adults. However, as winter temperatures rise, these mortality rates are likely to drop, leading to larger aphid populations in the spring.

Warmer winters mean that more aphids survive the cold months, ready to burst forth and colonize plants as soon as the weather warms up. This could lead to earlier and more severe infestations, as aphid populations start off larger and have more time to reproduce before the end of the growing season.

Case study: The cabbage aphid (*Brevicoryne brassicae*)

In a study conducted in the UK, researchers observed the effects of milder winters on cabbage aphid populations. Historically, harsh winters would significantly reduce their numbers, providing a reprieve for cabbage farmers in the spring. However, with increasingly mild winters, cabbage aphid populations have been able to survive in greater numbers. The result? Devastating spring infestations that lead to significant crop losses and increased pest management costs for farmers.

Precipitation patterns: The great migration

Changes in precipitation patterns are another aspect of climate change that could significantly influence aphid behavior. Aphids are highly sensitive to moisture levels, and shifts in rainfall could affect their migration and feeding patterns.

1 **Increased Rainfall:** While heavy rains can sometimes wash aphids off plants, prolonged periods of increased rainfall can lead to higher humidity levels that favor aphid reproduction and survival. This can result in rapid population growth and increased pressure on crops.
2 **Drought Conditions:** On the other hand, drought conditions can stress plants, making them more susceptible to aphid infestations. Stressed plants may produce more sap, providing an abundant food source for aphids and potentially leading to explosive population growth.

Case study: The pea aphid (*Acyrthosiphon pisum*) in drought conditions

During a prolonged drought in the Midwest United States, researchers observed a significant increase in pea aphid populations. The drought-stressed pea plants produced more sap, which not only attracted more aphids but also reduced the plants' ability to withstand the feeding damage. The result was a vicious cycle of increased aphid infestation and decreased plant health, leading to substantial yield losses.

Geographic range expansion: New frontiers for aphids

As temperatures rise, the geographic range of many aphid species is likely to expand. Areas that were previously too cold for aphids to survive may become new frontiers for these pests. This could lead to the spread of aphid-borne plant viruses and increased pressure on crops in regions that have not previously dealt with significant aphid problems.

Case study: The Russian wheat aphid (*Diuraphis noxia*) in Northern Europe

Historically confined to warmer regions, the Russian wheat aphid has been gradually expanding its range northward. With milder winters and warmer springs, this pest has started to establish populations in northern Europe, posing a new threat to wheat farmers in the region. The economic impact has been significant, with increased pest management costs and reduced yields as farmers grapple with this new invader.

Phenological shifts: The early bird gets the aphid

Climate change is also causing shifts in the timing of biological events, known as phenological shifts. For aphids, this means earlier emergence and longer growing seasons, providing more opportunities for reproduction and population growth.

1 **Earlier Spring Emergence:** Warmer temperatures can cause aphids to emerge from their overwintering stages earlier in the spring. This early start allows them to get a head start on reproduction, leading to larger populations by mid-season.
2 **Extended Growing Seasons:** Longer growing seasons provide aphids with more time to complete multiple generations. In regions with mild autumns, aphids can continue reproducing later into the year, further compounding their population growth.

Case study: The green peach aphid (*Myzus persicae*) in extended growing seasons

In California, the green peach aphid has benefited from extended growing seasons, thanks to milder autumns and earlier springs. Farmers have observed that these aphids are now able to complete more generations per year, leading to more severe infestations and increased difficulty in managing their populations. The economic impact has been felt through higher pest control costs and reduced crop quality.

The future of pest management in a changing climate

Given the potential for climate change to create an aphid utopia, it is essential to consider how pest management strategies will need to evolve to keep up with these changes. Here are a few potential directions:

1 **Climate-Adaptive Integrated Pest Management (IPM):** Developing IPM strategies that are adaptive to changing climate conditions will be crucial. This might include monitoring climate trends, adjusting planting schedules, and using predictive models to anticipate aphid outbreaks.
2 **Enhanced Biological Controls:** Leveraging natural predators and biocontrol agents that can adapt to changing climate conditions will be essential. This could involve breeding or selecting biocontrol agents that thrive in warmer and more variable climates.
3 **Resilient Crop Varieties:** Developing crop varieties that are more resistant to aphid infestations and climate stress will be a key focus of agricultural research. This might involve traditional breeding techniques as well as advanced genetic engineering.

4　**Technological Innovations:** Employing advanced technologies, such as drones, AI, and precision agriculture, to monitor and manage aphid populations in real-time will become increasingly important. These technologies can help farmers respond more quickly and effectively to aphid threats.

5　**Policy and Collaboration:** Governments, researchers, and farmers will need to collaborate closely to develop policies and strategies that address the challenges posed by climate change and aphid infestations. International cooperation will be vital to share knowledge, resources, and best practices.

The role of biodiversity

Maintaining biodiversity in agricultural ecosystems is not just a noble environmental goal—it's a practical strategy to enhance resilience against aphid outbreaks. Diverse plantings can disrupt the lifecycle of aphids, create less hospitable environments for them, and provide habitats for their natural predators. By fostering a more complex ecosystem, farmers can leverage biodiversity to naturally regulate pest populations, including aphids.

Biodiversity as a natural defense mechanism

Biodiversity in agriculture means incorporating a variety of plants, trees, and even animals into the farming system. This can include crop rotations, intercropping, planting cover crops, and maintaining hedgerows and wildflower borders. Each of these elements plays a role in creating a robust ecosystem that can resist pest invasions more effectively than monocultures.

Improving Biodiversity in Agriculture Systems

1　**Crop Rotation:** Rotating crops from one season to the next prevents aphids from finding their preferred host plants year-round, breaking their lifecycle and reducing the chances of large infestations.

2　**Intercropping:** Planting different crops together can confuse and repel aphids. Some plants emit chemicals or scents that deter aphids or attract their predators, creating a natural pest control system.

3　**Cover Crops:** These are plants grown primarily to benefit the soil and ecosystem rather than for harvest. They can provide habitat for beneficial insects that prey on aphids, and their roots can improve soil health, making it less hospitable for aphid survival.

4　**Hedgerows and Wildflower Borders:** These areas serve as refuges and hunting grounds for beneficial insects like ladybugs, lacewings, and hoverflies, all of which are natural aphid predators. By maintaining these habitats, farmers support a thriving population of natural pest controllers.

Case study: Biodiversity buffers

A study conducted in the United Kingdom provided compelling evidence for the effectiveness of biodiversity in controlling aphid populations. Researchers examined farms with diverse crop rotations and intercropping practices and compared them to monoculture farms. The results were striking: farms that embraced biodiversity had significantly lower aphid infestations.

The increased presence of natural predators supported by a variety of plants, played a crucial role in keeping aphid populations in check. For instance, hoverflies, whose larvae feed on aphids, were more abundant on farms with a mix of crops and flowering plants. These predators could easily move between plants, finding aphid colonies and reducing their numbers naturally.

Key findings from the study

- **Reduced Aphid Infestations:** Farms with diverse planting practices saw up to a 40% reduction in aphid populations compared to monoculture farms.
- **Increased Predator Presence:** There was a noticeable increase in the populations of natural aphid predators like ladybugs and hoverflies on diversified farms.
- **Enhanced Crop Health:** Crops on farms with high biodiversity exhibited fewer signs of aphid damage and overall better health, likely due to the combined effects of reduced pest pressure and improved soil conditions.

Practical implementation of biodiversity

Implementing biodiversity on farms requires thoughtful planning and commitment, but the benefits are substantial. Here are some practical steps farmers can take:

1 **Diversified Planting Plans:** Develop a planting schedule that includes rotating different crops each season and incorporating intercropping strategies.
2 **Planting Hedgerows and Wildflowers:** Establish these around field borders and within crop areas to provide habitat for beneficial insects.
3 **Using Cover Crops:** Integrate cover crops into the farming system during off-seasons to improve soil health and provide continuous habitat for predators.
4 **Encouraging Wildlife:** Create ponds or small water features, birdhouses, and insect hotels to attract and support a variety of wildlife that can help with pest control.

The economic benefits of biodiversity

While implementing biodiversity practices might require an initial investment of time and resources, the long-term economic benefits can be significant. By reducing reliance on chemical pesticides, farmers can lower their input costs and minimize the risk of developing pesticide-resistant aphid populations. Healthier crops and higher yields, resulting from reduced pest pressure and improved soil health, translate into better financial returns.

Moreover, farms that prioritize biodiversity often benefit from ecosystem services such as pollination, water regulation, and soil fertility, all of which contribute to sustainable agricultural productivity.

Case study: Economic analysis of biodiverse farming

In a follow-up to the UK study, researchers conducted an economic analysis comparing the costs and benefits of biodiversity-friendly practices. They found that: Farmers practicing crop rotation and intercropping spent 30% less on chemical pesticides. These farmers also reported a 20% increase in crop yields due to better pest control and healthier soil. Some farmers were able to market their produce as organic or sustainably grown, commanding higher prices.

Future research directions

Artificial intelligence and big data

Harnessing the power of AI and big data could revolutionize our approach to pest management. By analyzing vast datasets on weather patterns, aphid behavior, and crop health, AI can predict aphid outbreaks and suggest timely interventions. This technology offers a proactive approach to pest control, moving away from reactive measures to more precise, preemptive strategies.

AI in Action Imagine a future where farmers receive real-time alerts on their smartphones, advising them on the best time to apply biopesticides or release natural predators, all based on predictive algorithms. This proactive approach could significantly reduce crop losses and improve pest management efficiency.

AI and big data can help farmers make data-driven decisions, optimizing the timing and methods of pest control. For instance, an AI system could integrate data from weather stations, satellite imagery, and ground sensors to forecast aphid population spikes. Farmers would then receive tailored advice on how to protect their crops, reducing the guesswork and increasing the effectiveness of their pest management efforts.

Example: Predictive pest management platform

An example of this technology is a predictive pest management platform developed by researchers at a leading agricultural university. This platform uses machine learning algorithms to analyze historical and real-time data, providing farmers with predictive insights into pest pressures. The system can recommend specific actions, such as the optimal timing for biopesticide application or the introduction of biological controls, based on the predicted aphid activity.

Advanced biopesticides

Research into biopesticides is progressing at a rapid pace. Scientists are developing new formulations that are highly specific to aphids, minimizing harm to beneficial insects and the environment. These biopesticides are derived from natural sources, such as bacteria, fungi, and plant extracts, and they target aphids without the broad-spectrum impact of traditional chemical pesticides.

Case Study: The Biopesticide Breakthrough A recent breakthrough in biopesticide research led to the development of a fungal pathogen that specifically targets aphids. Field trials in Brazil showed a dramatic reduction in aphid populations with minimal impact on other insect species. This innovation holds promise for a more targeted and sustainable approach to pest control.

Benefits of Advanced Biopesticides

- **Targeted Action:** Specifically designed to affect aphids, reducing non-target impacts.
- **Environmental Safety:** Lower environmental impact compared to conventional pesticides.
- **Resistance Management:** Reduced likelihood of pests developing resistance.
- **Sustainability:** Supports long-term ecological balance.

Researchers are also exploring the potential of using genetically engineered microorganisms to deliver biopesticides directly to aphids. These microorganisms could be engineered to produce aphid-specific toxins or disrupt the aphid's reproductive processes, providing a novel method of pest control that is both effective and environmentally friendly.

Genetically modified crops

As genetic engineering technologies advance, we may soon see crops that are entirely resistant to aphids. By tweaking the DNA of plants to produce natural aphid repellents or toxins, scientists aim to create a future where aphid infestations are a thing of the past.

> ## GM Crops in Pest Management
> - **Bt Crops:** Genetically modified to produce Bacillus thuringiensis (Bt) toxin, which is harmful to certain pests but safe for humans and non-target species.
> - **Herbicide Resistance:** Allows crops to survive applications of herbicides that kill weeds, reducing competition for resources.
> - **Nutrient Fortification:** Enhances the nutritional profile of crops, benefiting human health.

One promising avenue is the development of crops that produce aphid-specific pheromones or other signaling compounds that disrupt aphid behavior. These plants could repel aphids or attract natural predators, effectively using the aphids' own biology against them. Such innovations could significantly reduce the need for chemical interventions, promoting a more sustainable approach to agriculture.

The future of pest management

Integrated pest management (IPM) 2.0

The future of pest management lies in the integration of multiple strategies, combining the best of traditional methods with cutting-edge technology. IPM 2.0 will likely involve a mix of genetic engineering, biological controls, precision agriculture, and AI-driven decision-making.

> ## The Pillars of IPM 2.0
> 1. **Genetic Engineering:** Developing pest-resistant crop varieties.
> 2. **Biological Controls:** Using natural predators and biopesticides.
> 3. **Precision Agriculture:** Employing technology for targeted interventions.
> 4. **AI and Big Data:** Predictive analytics for proactive pest management.

Genetic engineering

The advent of genetic engineering has opened new doors in pest management. Scientists are now able to develop crop varieties that are inherently resistant to pests like aphids. These genetically modified crops can produce natural repellents or toxins that deter pests, reducing the need for chemical pesticides.

Biological controls

Using nature to combat nature, biological controls involve introducing natural predators or parasites to control pest populations. Ladybugs, lacewings, and parasitoid wasps are just a few examples of beneficial insects that can help keep aphid populations in check. Additionally, advancements in biopesticides—derived from natural materials such as bacteria and fungi—offer targeted pest control with minimal environmental impact.

The role of policy and education

Effective pest management requires support from policymakers and education for farmers. Governments must invest in research, provide subsidies for sustainable practices, and ensure that farmers have access to the latest knowledge and technologies.

Policy success in Europe

The European Union's stringent regulations on pesticide use have spurred innovation in alternative pest control methods. By providing grants for research and development, the European Union has encouraged the adoption of IPM practices, leading to a decline in chemical pesticide use and an increase in sustainable farming methods.

> **European Success Story**
>
> In France, a government-funded program helped farmers transition to organic farming methods. By offering financial incentives and technical support, the program reduced pesticide use by 30% and increased biodiversity in agricultural landscapes.

Education and training

Educating farmers about the benefits and implementation of IPM 2.0 is crucial. Extension services and agricultural organizations play a vital role in disseminating information and training farmers. Workshops, field demonstrations, and online courses can equip farmers with the knowledge and skills needed to adopt advanced pest management strategies.

Online resources: Research collaborations and opportunities

For those eager to dive deeper into the future of aphid management, numerous online resources are available. From research databases to collaborative platforms, these tools can provide valuable insights and foster innovation.

Online resource hub

- **Research Databases:** Access the latest studies on aphid biology, pest control methods, and environmental impacts.
- **Collaborative Platforms:** Connect with researchers, farmers, and policymakers to share knowledge and develop new solutions.
- **Educational Materials:** Find webinars, tutorials, and guides on implementing advanced pest management strategies.

Conclusion

As we peer into the future, it becomes clear that the battle against aphids is far from over. However, with the advent of new technologies, innovative research, and a commitment to sustainable practices, we can envision a future where aphid populations are managed more effectively and with less environmental impact.

By embracing the possibilities offered by genetic engineering, AI, and IPM, we can protect our crops, secure our food supply, and reduce the economic burden of aphid infestations. As we continue to push the boundaries of science and technology, the future of pest management looks brighter than ever.

Glossary of Terms

Aphid-Resistant Plants: Crop varieties that possess natural defenses against aphid infestation.

Biological Control: The use of natural predators or parasites to manage pest populations.

Chemical Insecticides: Synthetic chemicals used to kill or repel insects.

Integrated Pest Management (IPM): A sustainable approach to managing pests that combines biological, chemical, and cultural methods.

Parthenogenesis: A form of asexual reproduction where an organism produces offspring without fertilization.

Precision Agriculture: Farming practices that use technology to monitor and manage crops with high precision.

Sustainable Agriculture: Farming practices that maintain and improve environmental health, economic profitability, and social equity.

Bibliography

Blackman, R. L., & Eastop, V. F. (2000). *Aphids on the World's Crops: An Identification and Information Guide.* Wiley.

Dixon, A. F. G. (1998). *Aphid Ecology: An Optimization Approach.* Chapman & Hall.

Makkouk, K. M., & Kumari, S. G. (2009). *Viruses infecting pepper crops. Advances in Virus Research,* 75, 1–35.

Simon, J. C., Rispe, C., & Sunnucks, P. (2002). *Ecology and evolution of sex in aphids. Trends in Ecology & Evolution,* 17(1), 34–39.

Van Emden, H. F., & Harrington, R. (2007). *Aphids as Crop Pests.* CABI.

CHAPTER 10

Conclusion: A call to arms (and legs)

Introduction

Picture this: It's a sunny day in your garden, the tomatoes are ripening, the roses are in full bloom, and everything seems perfect. But then, like a bad horror movie, you see them—tiny, green invaders sucking the life out of your precious plants. Yes, aphids. Those persistent little bugs can turn any gardener's dream into a nightmare.

But before you reach for the pesticide, take a moment to appreciate the sheer tenacity and adaptability of these tiny creatures. Aphids, despite their minuscule size, are masters of survival. They've been around for millions of years, adapting to every environmental challenge thrown their way. In this final chapter, let's take a step back and reflect on what aphids have taught us and how we can learn to coexist with them.

The invincible aphid: A lesson in resilience

Aphids are the ultimate survivors. Their ability to reproduce rapidly and adapt to changing conditions makes them a formidable foe. But it also makes them a fascinating subject of study. Aphids remind us that resilience isn't about being the biggest or the strongest—it's about adaptability, persistence, and the ability to thrive in the face of adversity.

Think about it. These tiny insects have developed ways to reproduce without males, form mutualistic relationships with ants, and resist many of our chemical attacks. They've outsmarted us at almost every turn, proving that size doesn't matter when it comes to survival.

DOI: 10.1201/9781003537786-10

Figure 10.1 Aphid mummy. (Shutterstock ID: 2177849503)

Learning from aphids: Humor and humility

It's easy to get frustrated with aphids, but maybe it's time to take a leaf out of their book—pun intended. Aphids teach us the importance of adaptability and resilience. They've shown us that survival often requires flexibility and a willingness to change. And perhaps, most importantly, they've reminded us that even the smallest creatures can have a significant impact on our world.

So, let's approach our battle with aphids with a sense of humor and humility. After all, we're not just fighting pests—we're engaging with one of nature's most persistent success stories.

Final Thoughts on Aphid Management

1 **Stay Informed:** Knowledge is your best weapon. Stay updated on the latest research and management strategies.
2 **Be Patient:** Pest management is an ongoing process. Patience and persistence are key.
3 **Think Holistically:** Integrated pest management (IPM) combines multiple strategies for more effective and sustainable control.
4 **Embrace Natural Solutions:** Encourage beneficial insects and use organic methods to maintain a balanced ecosystem.

Embracing a balanced approach

Instead of waging an all-out war on aphids, consider a more balanced approach. IPM offers a sustainable way to control aphid populations while minimizing harm to the environment. This approach combines biological, cultural, and chemical methods, ensuring that we protect our crops and our planet.

Encouraging beneficial insects, using insecticidal soaps, and planting aphid-resistant varieties are all part of this holistic strategy. By integrating these methods, we can manage aphid populations more effectively and sustainably.

A call to arms (and legs)

Now, as we wrap up our journey through the world of aphids, it's time to call on all gardeners, farmers, and nature enthusiasts to take up the challenge. Let's embrace the lessons aphids have taught us about resilience and adaptability. Let's work together to develop sustainable pest management strategies that protect our crops and respect the delicate balance of nature.

Most importantly, let's keep our sense of humor. After all, in the grand scheme of things, aphids are just tiny bugs trying to make their way in the world. They're not evil masterminds plotting our downfall—they're simply doing what they need to survive.

Inspirational Quotes on Resilience:

- "In the confrontation between the stream and the rock, the stream always wins—not through strength, but through persistence." —H. Jackson Brown Jr.
- "It is not the strongest of the species that survive, nor the most intelligent, but the one most responsive to change." —Charles Darwin
- "Our greatest glory is not in never falling, but in rising every time we fall." —Confucius

---

Looking ahead: Innovation and collaboration

As we look to the future, it's clear that innovation and collaboration will be key to managing aphid populations. Advances in genetic engineering, AI-driven pest detection, and precision agriculture offer exciting possibilities for more effective and sustainable pest management.

But no matter how advanced our technology becomes, the basic principles of patience, persistence, and balance will remain essential. By embracing these principles and working together, we can protect our crops, secure our food supply, and reduce the economic burden of aphid infestations.

The role of community

Managing aphid populations isn't just a task for individual farmers and gardeners—it's a community effort. By sharing knowledge, resources, and experiences, we can develop more effective strategies and support each other in our efforts to maintain healthy gardens and farms.

So, let's come together as a community. Let's learn from each other, share our successes and failures, and work towards a future where we coexist with aphids in a way that benefits both us and the environment.

Top Innovations in Aphid Management:

1 **Biopesticides:** Environmentally friendly pesticides that target specific pests.
2 **Genetic Engineering:** Creating pest-resistant crop varieties.
3 **AI and Big Data:** Predicting pest outbreaks and optimizing interventions.
4 **Precision Agriculture:** Using technology to monitor and manage crops more effectively.

Final thoughts

As we conclude this journey through the world of aphids, let's remember that these tiny insects, despite their size, have a significant impact on our gardens, farms, and ecosystems. They challenge us to be better stewards of the land, to innovate and adapt, and to find balance in our interactions with nature.

So, here's to a future where we coexist with aphids, learning from their tenacity and finding new ways to thrive alongside them. Let's approach our battle with aphids with creativity, resilience, and a good dose of humor. Together, we can develop sustainable strategies that protect our crops and gardens while respecting the intricate web of life that aphids are a part of.

Glossary of Terms

Aphid-Resistant Plants: Crop varieties that possess natural defenses against aphid infestation.

Biopesticides: Pesticides derived from natural materials such as animals, plants, bacteria, and minerals.

Genetic Engineering: The direct manipulation of an organism's genes using biotechnology.

IPM (Integrated Pest Management): A sustainable approach to managing pests that combines biological, chemical, and cultural methods.

Precision Agriculture: Farming practices that use technology to monitor and manage crops with high precision.

Sustainable Agriculture: Farming practices that maintain and improve environmental health, economic profitability, and social equity.

Bibliography

Blackman, R. L., & Eastop, V. F. (2000). *Aphids on the World's Crops: An Identification and Information Guide*. Wiley.

Dixon, A. F. G. (1998). *Aphid Ecology: An Optimization Approach*. Chapman & Hall.

Makkouk, K. M., & Kumari, S. G. (2009). Viruses infecting pepper crops. *Advances in Virus Research, 75*, 1–35.

Raupp, M. J., & Shrewsbury, P. M. (2009). Sustainable pest management in urban landscapes. *Annual Review of Entomology, 54*, 233–250.

Simon, J. C., Rispe, C., & Sunnucks, P. (2002). Ecology and evolution of sex in aphids. *Trends in Ecology & Evolution, 17*(1), 34–39.

Van Emden, H. F., & Harrington, R. (2007). *Aphids as Crop Pests*. CABI.

Bibliography

Albrecht, D. (1997). *The Great Piece of Turf*. Nuremberg: Engraving. Renaissance artistry in botanical illustration.

Atwood, M. (1985). *The Handmaid's Tale*. Toronto: McClelland and Stewart. Symbolism of aphids in dystopian literature.

Blackman, R. L., & Eastop, V. F. (2007). *Taxonomic issues and insect-plant interactions in aphids*. European Journal of Entomology, 104(3), 267–282. https://doi.org/10.14411/eje.2007.037

Braendle, C., Davis, G. K., Brisson, J. A., & Stern, D. L. (2006). *Wing dimorphism in aphids*. Heredity, 97(3), 192–199. https://doi.org/10.1038/sj.hdy.6800863

Brisson, J. A. (2010). *Aphid wing dimorphisms: Linking environmental and genetic control of trait variation*. Philosophical Transactions of the Royal Society B: Biological Sciences, 365(1540), 605–616. https://doi.org/10.1098/rstb.2009.0255

BugLife. (2015). *Environmental Installations Highlighting Aphid Roles*. Art installations emphasizing ecosystem balance.

Caillaud, M. C., & Via, S. (2000). *Specialized feeding behavior influences both ecological specialization and assortative mating in sympatric host races of pea aphids*. The American Naturalist, 156(6), 606–621. https://doi.org/10.1086/316992

Carson, R. (1962). *Silent Spring*. Boston: Houghton Mifflin. Ecological imbalances caused by pesticide use on aphid populations.

Christie, A. (1926). *The Murder of Roger Ackroyd*. London: William Collins & Sons. Aphid-infested garden as a metaphor for hidden decay.

Dedryver, C. A., Le Ralec, A., & Fabre, F. (2010). *The conflicting relationships between aphids and men: A review of aphid damage and control strategies*. Comptes Rendus Biologies, 333(6–7), 539–553. https://doi.org/10.1016/j.crvi.2010.03.009

Dixon, A. F. G. (1998). *Aphid Ecology: An Optimization Approach*. Springer.

Dixon, A. F. G., & Kindlmann, P. (1999). *Cost of flight apparatus and optimum body size of aphid migrants*. Ecology, 80(5), 1678–1690. https://doi.org/10.1890/0012-9658(1999)080[1678:COFAAO]2.0.CO;2

Donaldson, J. (2009). *What the Ladybird Heard*. London: Macmillan Children's Books. Depiction of aphids in children's literature.

Fenton, B., Malloch, G., & Woodford, J. A. T. (1998). *Analysis of clonal diversity and genetic differentiation of the peach–potato aphid, Myzus persicae (Sulzer), using microsatellites*. Molecular Ecology, 7(12), 1405–1411.

Foster, W. A., & Northcott, A. D. (1994). *The role of soldier aphids in the defense of aphid colonies (Homoptera: Pemphigidae)*. Oecologia, 100(2), 140–145. https://doi.org/10.1007/BF00317138

Hales, D. F., Wilson, A. C. C., & Simon, J. C. (2014). *Aphids: Clones in space*. In D. Bickel & P. Cranston (Eds.), *Annual Review of Entomology* (Vol. 59, pp. 547–567).

Jander, G., & Howe, G. A. (2008). *Plant interactions with aphid herbivores: Molecular and ecological perspectives*. Current Opinion in Plant Biology, 11(4), 420–426. https://doi.org/10.1016/j.pbi.2008.05.003

Kingsolver, B. (2012). *Flight Behavior*. New York: Harper. Ecological themes reflected through aphid activity.

Le Trionnaire, G., Hardie, J., Jaubert-Possamai, S., Simon, J. C., & Tagu, D. (2008). *Shifting from clonal to sexual reproduction in aphids: Physiological and developmental aspects*. Biology of Reproduction, 78(5), 507–516.

Moran, N. A. (1992). *The evolution of aphid life cycles*. Annual Review of Entomology, 37(1), 321–348. https://doi.org/10.1146/annurev.en.37.010192.001541

Moran, N. A., & Telang, A. (1998). *Buchnera aphidicola and other symbionts of aphids provide nutrients to their hosts*. Entomologia Experimentalis et Applicata, 86(3), 233–236. https://doi.org/10.1046/j.1570-7458.1998.00298.x

Orwell, G. (1945). *Animal Farm*. London: Secker and Warburg. Aphids as symbols of oppression in political allegories.

Pérez-Hedo, M., Bouagga, S., Jaques, J. A., & Urbaneja, A. (2015). *The predatory mirid Nesidiocoris tenuis as a biological control agent: A review of its effects on pests and non-target arthropods*. Pest Management Science, 71(6), 822–830. https://doi.org/10.1002/ps.3991

Roitberg, B. D., & Myers, J. H. (1979). *Behavioral and physiological adaptations of pea aphids (Homoptera: Aphididae) to high densities*. Canadian Journal of Zoology, 57(6), 1117–1123. https://doi.org/10.1139/z79-144

Simon, J. C., Rispe, C., & Sunnucks, P. (2002). *Ecology and evolution of sex in aphids*. Trends in Ecology & Evolution, 17(1), 34–39. https://doi.org/10.1016/S0169-5347(01)02331-X

Stern, D. L., & Foster, W. A. (1996). *The evolution of soldiers in aphids*. Biological Reviews, 71(1), 27–79.

Van Emden, H. F., & Harrington, R. (Eds.). (2007). *Aphids as Crop Pests*. London: CABI Publishing. https://doi.org/10.1079/9780851998190.0000

Vorburger, C., & Rouchet, R. (2016). *Are aphid parasitoids locally adapted to the prevalence of defensive symbionts in their hosts?* BMC Evolutionary Biology, 16(1), 271.

Index